T0092266

INSIDE THE WORLD OF CLIMATE CHANGE SKEPTICS

Kristin Haltinner &
Dilshani Sarathchandra

INSIDE THE WORLD OF CLIMATE CHANGE SKEPTICS

University of
Washington Press
Seattle

Inside the World of Climate Change Skeptics was supported by a grant from the McLellan Endowment, established through the generosity of Martha McCleary McLellan and Mary McLellan Williams.

Design by Mindy Basinger Hill
Composed in Adobe Caslon Pro and Source Sans Pro

UNIVERSITY OF WASHINGTON PRESS
uwapress.uw.edu

Cataloging information is available from the Library of Congress
LCCN 2023932625
ISBN 9780295751290 (hardcover)
ISBN 9780295751306 (paperback)
ISBN 9780295751313 (ebook)

♾ This paper meets the requirements of ANSI/NISO Z39.48-1992 (Permanence of Paper).

To Ethan, Young,
and Braxton.
To a thriving planet
for you and
all living beings.

KRISTIN
HALTINNER

To Deelaka,
Chathura, Kiara,
and Kian,
who will see
the future.

DILSHANI
SARATHCHANDRA

contents

acknowledgments

WE WOULD LIKE TO RECOGNIZE the phenomenal community that supported us in this work and the creation of this manuscript. To the administrators who supported this work, Dr. Sean Quinlan, Dr. Traci Craig, and Dr. Brian Wolf. To our amazing colleagues—the best in the world—Dr. Leontina Hormel, Dr. Ryanne Pilgeram, and Dr. Deborah Thorne. To our collaborators, Dr. Steven Radil, Dr. Jennifer K. Ladino, and Dr. Tom Ptak.

We extend our deep gratitude to the journals that published our related work, along with the journal editors and manuscript reviewers, whose feedback has strengthened our work: *Sociology Compass*, *Public Understanding of Science*, *Contexts*, *Sociological Inquiry*, *Environmental Sociology*, *Rural Sociology*, *Climate*, *Global Environmental Change*, *Social Currents*, *Emotion, Space and Society*, *Journal of Rural Social Sciences*, and *Environment: Science and Policy for Sustainable Development*.

Thank you to our amazing team of research assistants, Ashli George, Sarah Olsen, Kathryn Pawelko, Christine Sedgwick, Hannah Spear, Randolph P. Stuart, and Amber Ziegler. This would not have happened without your dedication, reliability, and thoughtfulness. Gratitude to Dr. Matthew Grindal, who helped us shape our assessment of the climate skeptic identity.

To our families, who supported our commitment to this project. Dennis and Joanna Haltinner, without whom Kristin's career would not have happened. To Douglas and David Haltinner and Micah Darling, for their love, encouragement, and occasional antagonism. To Garrett, Young, and Ethan Clevinner, who provided much needed breaks but also the drive to persist. To Malani Gamage, Upali Sarathchandra, Chamara Sarathchandra, Renuka Dayanthi, Amarasuriya Wijesekara, and Udith Kolambage, for being Dilshani's biggest cheerleaders over the years. To her friends Callista

Rakhmatov, Carey Wilson, and Cedric Taylor, whose support and insights have helped her in both work and life.

Finally, to Andrew Berzanskis, our wonderful editor, who had utmost patience with us through the process and is perhaps the world's fastest email responder.

INSIDE THE WORLD OF CLIMATE CHANGE SKEPTICS

introduction

DOUGLAS IS AN ELEMENTARY SCHOOL biology teacher who does not believe in anthropogenic (human-caused) climate change. He has "looked at" climate science data and associated findings and has "uncovered problems" with its methodology and conclusions. As a result of his personal "research," Douglas rejects evidence that storms have been increasing in intensity; he argues that we need better data to conclude that carbon dioxide levels have risen over the decades. In fact, he contends that what scientists mistake as sea level rise is really the land sinking.

Douglas is among the 30 percent of Americans who do not believe that climate change is caused by human activities. Actually, according to Yale Climate Opinion maps (2021), he is part of an even smaller group, the 14 percent of Americans who reject that climate change is happening at all.[1]

Douglas identifies as an "apolitical Christian" who is "socially conservative" and an "environmentalist." Douglas has always cared about the natural world. Growing up in Appalachia, he spent a lot of time outdoors with his family. His primary passion is birds and preserving bird habitats. At the time of our interview he was very concerned about then-president Donald Trump's move to cut funding to the Environmental Protection Agency (EPA). He believes that protecting air and water quality and preserving wildlife habitat is "fundamental."

While Douglas's environmentalism and concurrent climate change skepticism may sound incongruous to those who accept the science around climate change, his beliefs mirror those of many of his peers. In fact, in our survey of one thousand self-declared climate change skeptics, we found that 87.7 percent agree that "even if climate change isn't real, we should work hard to preserve the environment."

Douglas and the other participants in our sample suggest that several environmental concerns and pro-environmental policies are broadly supported by most people (e.g., pollution, waste, habitat destruction, renewable energy) regardless of their specific perspectives on climate change. This has important and hopeful implications for journalists, scientists, and science writers as they communicate climate science and for politicians as they develop environmental policy.

Climate change skepticism is not simply the ignorance of a few individuals. Rather, it is an outcome of a concerted effort to sow misinformation and doubt. The so-called climate change denial countermovement began in earnest in the United States in the mid-1990s. It is not the first movement of its kind but is, rather, embedded within a broader antiscience, anti-intellectual movement (called the Reflexivity Countermovement by sociologists Aaron McCright and Riley Dunlap) that began decades earlier.[2] These movements have relied on right-leaning corporations and think tanks to advance a false narrative that rejects climate science and the veracity of climate change. They have falsely presented or withheld the results of climate science, harassed individual climate scientists, used political power to change previously accepted rules around scientific research, and bolstered the voices of a few contrarian scientists who reject climate change.[3] Through this process, conservative think tanks, members of the fossil fuel industry, conservative politicians, and others have followed what is often called the "tobacco model," putting massive amounts of money into promoting minority, contrarian scientific voices through conservative media.[4]

This massive countermovement has convinced a substantial portion of ideological conservatives that climate change is either a natural occurrence or not happening at all. While 94 percent of Democrats believe in the reality of anthropogenic climate change, only 69 percent of Republicans think that humans are contributing to global warming.[5]

Who, then, are these climate change skeptics? Work in sociology suggests that such skeptics are more likely to be political conservatives, men, and regular churchgoers who hold lower levels of environmentalism, compared to other Americans.[6] Of all of these factors, the most central is political orientation.[7] Political ideology matters to such a degree that it influences other things like education's impact on climate perceptions: those who are politically conservative tend to hold more skeptical beliefs about climate change regardless of their level of education.[8] This skepticism is reinforced among these populations through cognitive biases such as confirmation bias (the tendency to reject information that challenges one's existing beliefs), motivated reasoning (that leads to the rejection of authority and experts), and associated conspiracy ideation.[9]

Existing research presents various typologies of climate change skeptics and skepticism. For example, ocean scientist Stefan Rahmstorf identified three types: "trend sceptics (who deny there is global warming), attribution sceptics (who accept the global warming trend but see natural causes for this), and impact sceptics (who think global warming is harmless or even beneficial)."[10] Subsequent studies have added a fourth category, "consensus skeptics," who believe that there is no strong agreement among scientists about the reality and human causes of climate change.[11]

While recognizing these trends, this book attempts to provide greater insight about the nature of climate change skepticism. Rather than comparing skeptics to nonskeptics, as previous scholars have done, we draw from a combination of rich interview data and surveys among self-identified climate change skeptics (and some former skeptics) to present a deep exploration of the diversity of opinions, beliefs, and behaviors of those who do not accept climate science. We find that many skeptics experience a shared identity: they believe that they are being marginalized by society and prevented from participating in public discussion on complex issues. We also uncover that skeptics hold complicated and varied beliefs regarding science and the role of God as it relates to the climate. These complex positions further shape and are impacted by skeptics' engagement with information (e.g., news stories and scientific knowledge), which has implications for their emotional responses to climate change and support for environmental policy.

It is important to pause here and clearly define the terms we use. Social scientists have wrestled with how to understand the variation in climate change skepticism (which ranges from being unsure about climate change to actively and obstinately denying it). Prior scholarship employs competing language to define people who fail to accept climate science. Sociologists Dunlap and McCright refer to people who reject climate science as "deniers." Others, such as Rahmstorf, employ the word "skeptic." Drawing on our work published in the journal *Global Environmental Change*, we use "skeptic" and "skepticism" as umbrella terms to describe people who actively reject climate science *and* people who are uncertain about climate change.[12] We use the term "doubter" to refer to people who are unsure about climate change and "denier" for people who actively reject climate science. In chapter 7 we extend this characterization to propose climate change skepticism as a "continuum," with extreme denial on one end and doubt/uncertainty on the other.

At a more fundamental level, we must also clarify our use of the terms "climate change" and "global warming." While the two terms are often used interchangeably per scientific convention, we employ the term "climate change" to refer to both human-caused and naturally produced warming of the planet and the resulting effects.[13] When citing prior scholarship, surveys, and opinion polls that use the term "global warming," we retain that term. When quoting or paraphrasing research participants' perspectives, we retain the terms they use.

Our work makes an important contribution to the scholarship on climate change skepticism. By better understanding the complex nature of skepticism, social scientists, policymakers, and journalists will be better equipped to communicate climate science to skeptics. Communication scientist Matthew Nisbet argues that to "break through the communication barriers of human nature, partisan identity, and media fragmentation," we need to frame messages in ways that resonate with specific audiences.[14] By understanding the nature of climate change skepticism, the role of ideology and identity in skepticism, and the factors that contribute to people changing their minds, we can develop communication strategies that are able to break through these walls and filters. Policy initiatives, too, can be enriched through better understanding of skepticism (for example, addressing the fact that many skeptics hold pro-environmental views regarding habitat preservation, pollution, and renewable energy).[15] If implemented, these policies could make significant bipartisan improvements to climate preservation.

OUR DATA AND METHODOLOGY

To understand the intricacies of climate change skepticism we spoke directly with people who hold a variety of beliefs about climate science. In this book we rely on three original data sets: thirty-three interviews with self-identified climate change skeptics in Idaho; one thousand surveys with climate change skeptics in the US Pacific Northwest; and twenty-one interviews with people in Idaho who have changed their minds about climate change.[16]

Between May 2017 and December 2018 we conducted thirty-three interviews with self-identified climate change skeptics, each lasting between 30 and 120 minutes. We asked participants about their perspectives on climate change, additional questions about a variety of environmental topics, and their past and current relationship to science, scientists, media, and information sources. Details on demographic backgrounds of these interview participants are available in the appendix, table 1. Findings from these first-round interviews are featured in chapters 1 to 7.

Once we completed the analysis of the first-round interview data, we pored over our findings and existing scholarship to develop a survey to assess the perspectives of a broader swath of climate change skeptics. This survey was administered to one thousand adults living in Washington, Oregon, and Idaho, with assistance from Qualtrics, a firm that specializes in conducting representative online surveys. Data collection occurred between November 2019 and January 2020. Two initial screening questions led respondents to report whether they believe "climate change is happening" and "climate change is caused by human activities" (response categories: yes=1; no=2; not sure=3). Respondents who said that they believed climate change was happening and it is caused by human activities (in other words, respondents who answered yes to both screening questions) were screened out, limiting the resulting sample to only those who expressed denial or uncertainty regarding anthropogenic climate change. Participants who met this screening criteria then were asked to answer the full survey, which consisted of questions on beliefs regarding climate change, the environment, science and scientists, information sources, energy policy preferences, their emotional response to climate change, and demographic factors. Information on the demographic characteristics of the full survey sample is given in the appendix, table 2. Findings from the survey are interspersed throughout chapters 2 to 7, either directly or in reference to our previously published articles.

The final round of work consisted of interviews of twenty-one people who self-identified as having changed their minds about climate change.

In these interviews we sought to understand the factors that contributed to changing minds on controversial and heavily politicized issues such as climate change. Interviews began at the end of 2018 and lasted through the start of 2020. We asked participants about their early beliefs about climate change, the period when they changed their minds about climate change, and the nature and nuance of their current beliefs. These interview findings are featured in chapter 8, and additional information about the demographic characteristics of participants is in the appendix, table 1.

While we used standard statistical techniques to analyze survey data, our interview data analysis consisted of verbatim transcription of interviews, development of a coding system to identify patterns that emerged, and theory-building using what social scientists call "inductive analysis." Throughout the book we use "grounded theory" as intended, applying it to a topical area (i.e., climate change skepticism), which requires further nuanced exploration to build theory.[17] In this way we are able to draw on the best of what both qualitative and quantitative data and analysis have to offer: assessing trends (who, what, where) and exploring the how and why of skepticism.

IDAHO AND THE PACIFIC NORTHWEST

Idaho is an ideal location to use for our qualitative research as it has a higher-than-average percentage of skeptics (36 percent of people in the state do not believe in anthropogenic climate change, vs. 30 percent in the nation); this leads to a greater possibility of finding nuance within skepticism.[18] According to the CATO institute, Idaho has the fourth-highest percentage of libertarians in the country, along with a large population of both evangelical Protestants (21 percent) and Latter-day Saints (19 percent).[19] These distinct populations allow us to capture people who approach skepticism from a variety of positions.

Our survey work branched out further to include skeptics in Oregon and Washington as well. This region hosts shared sociocultural traits often unknown to residents outside the Pacific Northwest. For instance, the

Center for the Study of the Pacific Northwest at the University of Washington reports that Idaho, Oregon, and Washington share a connection and identification with salmon and antipathy toward Californians. Residents of the Pacific Northwest also celebrate their low population density and widespread green spaces, which they view as antithetical to California's "polluted" and "overcrowded" environment.[20]

Despite these shared aspects of culture, the states' social values vary in significant ways. Eastern Washington, Oregon, and Idaho are more conservative politically than the western portions of Oregon and Washington.[21] People in Idaho have higher levels of religiosity (thirty-third in the nation, compared to Oregon at thirty-ninth and Washington at forty-fourth) and more who are skeptical about climate change (20 percent, vs. 13 percent in Washington and 15 percent in Oregon) and its human causes (36 percent, vs. 28 percent in Washington and 30 percent in Oregon).[22] Despite these distinctions, our data suggests that skeptics across the region are similar along other demographic lines. For instance, we found no significant statistical differences between Idaho, Washington, and Oregon skeptic populations in terms of their gender, race, and political ideology distributions.

Washington, Oregon, and Idaho all have similar histories of settler colonialism, resource extraction, and contemporary conflicts between environmentalists and loggers and ranchers. Since white settlers arrived at what became known as the US Pacific Northwest, resource extraction has been an important part of economic growth in the region. Settler colonialism and resultant exploitative economic practices such as logging, fur trade, and mining served as the foundation for early non-Native economies and still play a significant role in each state.[23] These practices also did more than exploit the land; they were at the heart of colonial conquest in the region and the racist exploitation of Asian laborers in the late 1800s.[24]

Environmentalism through the second half of the twentieth century led to a drastic rise in tensions between environmentalists and loggers in the 1990s, a period that was so contentious it is known as the "Timber Wars": a never-ending battle between people committed to preserving the integrity of ancient forests and the animals who live there and people whose livelihoods depend on the logging industry.[25]

However, as timber mills closed, new housing developers moved into the once rural communities in the region. Our colleague and sociologist Ryanne Pilgeram explores this process in her book *Pushed Out*. Demonstrating a phenomenon she calls "rural gentrification," Pilgeram examines how the failing economic bases of former mill towns first led to impoverished small communities, which then were bought up by massive economic developers as "trendy second home locations."[26]

The continued battle between those who support economic development rooted in extraction on the one side and environmental conservationists on the other, is evidenced in the ongoing conflicts between cattle ranchers and government land agencies (i.e., the Bureau of Land Management, the US Fish and Wildlife Service, and the US Forest Service). The perhaps most well-known example was the 2016 standoff at the Malheur National Wildlife Refuge, when far-right militia members, led by Idaho resident Ammon Bundy, seized the headquarters of the Oregon refuge and occupied it for over a month, until their eventual arrest.[27] These activists sought state control of federally protected land. Similar conflicts have occurred in other communities across the Pacific Northwest and western United States more broadly. In 2022 Bundy ran for governor of Idaho as an independent.

In short, the region is an ideal microcosm for research about climate change skepticism. The three states share a history of conflict between people committed to resource extraction (miners, ranchers, loggers) and environmentalists. They also hold shared cultural values around rural life. Yet, despite these commonalities, Idaho, Washington, and Oregon also feature distinct political and religious climates, making them ideal for a nuanced study of climate change skepticism.

POSITIONALITY

In the social sciences, researchers often reflect on their own positions vis-à-vis their research participants, their topics of investigation, and other aspects of their work. Our personal experiences and the ideologies to which we adhere shape the types of questions we ask, the way we interpret the data, and the way we write our findings. Thus it is essential that we reflect on what each of us brings to our research in order to do good and fair work. Our identities and the cultural stories told about them also shape the interactions we have with participants who may treat us in certain ways because of our gender, race, age, size, or abilities. Here, too, we need to be reflective to assess the veracity of the information we have gathered.

Kristin Haltinner approached this project as a political sociologist interested in the operation of power and privilege. Her primary research has focused on right-wing political movements, such as the Tea Party Patriots and the Minuteman Civil Defense Corps. Her academic interest in climate change skepticism is strongly influenced by the increasing climate anxiety she experienced following the birth of her son. Haltinner grew up in a politically divided town in the Midwest and has long been interested in how and why people with similar backgrounds and experiences are able to construct and inhabit profoundly distinct realities.

Dilshani Sarathchandra's interest in this project stems from a background in the sociology of science and science communication. Her primary research has focused on decision-making processes in science and public attitudes toward controversial scientific topics. Sarathchandra grew up in Sri Lanka, a small island nation south of India, where conversations about climate change have moved toward mitigation, adaptation, and the need to address the unique biophysical and social vulnerabilities of those who bear the brunt of its effects. She brings to this work her personal experiences among vulnerable communities and an academic background in science studies.

THEORIZING CLIMATE CHANGE SKEPTICISM

As social scientists, we draw on the ideas and contributions of others to build and enhance a clear understanding of climate change skepticism. Owing this process to sociologists Barney Glaser and Anselm Strauss, we conduct what is termed "grounded theory." Glaser and Strauss, in their book *The Discovery of Grounded Theory*, outline a process of theory-building for qualitative work wherein scholars look for patterns in their data, develop theories from these patterns, and return to the work of other scholars to enhance and extend their analysis.[28]

We applied a grounded theory strategy in our analysis of climate change skepticism. After completing the first round of interviews, we spent significant time examining the patterns we saw in the data and assessed the degree to which climate change skeptics appear to hold a shared identity that they constructed in opposition to nonskeptics. Then we turned to literature on identity in the social sciences to better understand the phenomena we saw.

The field of social psychology sits at the intersection of sociology and psychology. It is interested in interrogating how both one's mental state and social contexts influence an individual's behavior. Social psychologists consider how social influences—both real and imagined—shape one's thoughts, feelings, and actions.[29]

Two theories of identity have emerged from these intersections within social psychology: identity theory and social identity theory. These two theories share significant overlap but have three distinct attributes: the perceived basis of identities, the understanding of identity salience, and the activation processes of identities.[30] Identity can be used as a categorization or identification process in both theories, but in social identity theory, identity is rooted in awareness of being part of a large group or collective. This awareness typically leads to a natural comparison to people in other groups, resulting in a clear and distinct "in-group" and a clear and distinct "out-group."[31] Identity theory sees identity as a self-categorization based on one's social roles and associated expectations.[32] The strength or salience of one's identity is another factor treated differently between the two theories. Social identity theory explores how and why a particular identity is primed in a particular context; identity theory examines the degree to which one or another identity is activated in a particular social situation.[33] Finally, the two theories depart in their understanding of the "cognitive processes" that underly personal experience. Social identity theory considers the concept of "depersonalization," wherein people's group membership becomes prototypical and leads to things like stereotyping, in-group biases, and out-group derogation.[34] Identity theory, in contrast, focuses on what scholars call "self-verification," or the ways people behave that are either consistent or inconsistent with their sense of self. This is typically thought of in terms of the performance of roles. Despite these various distinctions, identity scholars contend that there is much agreement among the two theories and that the two theories may (or should) work together to define what Jan Stets and Peter Burke call "a general theory of the self."[35]

Given its focus on social categorization and group identity, we find that social identity theory best fits our analysis of climate change skepticism. However, we agree with scholars advocating for greater unity across these fields and consideration of the way these theories, when combined, may present a better, more robust understanding of the self, of identity, and of the way they are impacted by both interpersonal interactions and broader social processes.[36]

Social identity theory is often credited to the early work of Henri Tajfel and his examination of the influence of social factors in racial bias and discrimination. Throughout the 1970s and 1980s the theory was developed further by scholars such as John Turner, Michael Hogg, Penelope Oakes, Dominic Abrams, and Howard Giles, among others.[37] At its heart, social identity theory considers how group membership (i.e., by race, political party, school community, etc.) shapes sense of self. The theory contends that a group's set of norms and values influence (and even dictate) the very thoughts, beliefs, attitudes, feelings, and behaviors of group members. Social identities lead people to "self-categorization" by which they label themselves as members of the group and develop a positive sense of themselves and other group members. This categorization system is often made through comparisons with one or more out-groups that are perceived through a series of less-favorable stereotypes. This sort of prototypical thinking often leads to a "depersonalization" process: people are seen as representative of the groups to which they are assigned rather than as individuals.

In the context of climate change skepticism, we demonstrate that skeptics construct their in-group identity against a perceived out-group of climate scientists. Skeptics construct their self-identity as a group of misunderstood truth-seekers who are undervalued and persecuted by society at large. In contrast, they construct climate scientists as having weak morals, being easily swayed by funding pressures to falsify data and reports, and as being exclusive, denying skeptics their full participation in public debate.

This identity construction serves as the crux of climate change skepticism, but it is not the full story. In this book we use social identity theory and other social and psychological research to examine the nuances of a climate identity. We demonstrate how things like religious ideation, adherence to conspiracy theories, environmentalism, scientific trust, engagement with media, and even emotions are shaped by climate identity but also, in turn, shape the manifestations of skeptic identity itself. In doing so we paint a robust understanding of climate change skepticism and its operation as a nuanced identity. This conceptualization, in turn, presents important implications for climate communication and climate policy.

WHAT TO EXPECT IN THIS BOOK

As previously mentioned, this book examines the complex perspectives and experiences of climate change skeptics, moving beyond existing survey data that compares skeptics and nonskeptics to better understand the intricacies and nuances of skepticism itself through a consideration of climate change skepticism as an identity.

In chapter 1 we consider climate change skepticism as an opinion-based and stigmatized social identity through an examination of existing research on identity formation (social identity theory—e.g., Tajfel and Turner) and its impact on adherents. We demonstrate how climate change skeptics create a self-identity rooted in perceived marginalization. Further, we explore the demographic factors that correlate with climate change skepticism (religion, age, race, gender, political ideology, education, conspiracy ideation, and so on) and address why these relationships exist.

Chapter 2 considers skeptics' relationship to climate science and scientists and explores the nuance of their perspectives through both personal narratives and statistical analysis of our survey data. We present a thorough analysis of climate scientists as an out-group and the resultant "us-versus-them" relationship that develops and enhances the sense of exclusion that skeptics feel when participating in public discourse on climate change.

In chapter 3 we examine the influence of underlying ideologies on climate change skepticism and identity. We explore how both religious ideology and conspiracy ideation (i.e., the belief that climate change is a hoax) produce meaningful skeptic subgroups. We highlight three groups that are distinct from other skeptics: religious stewards, religious anti-environmentalists, and conspiracy-driven skeptics. We explore these groups' distinct perspectives regarding trust in science, trust in media, environmentalism, and emotional responses to climate change.

Following this examination of the operation of ideology, in chapter 4 we explore the broader environmental beliefs of climate change skeptics and their associated environmental behaviors. Despite not believing in climate change, most of the people we talked with and surveyed do care about the environment. Specifically, skeptics point to pollution and habitat destruction as two environmental problems about which they have grave concerns and want to solve.

In chapter 5 we look at the sources from which skeptics access their information, which information sources they trust, their beliefs regarding how information should be presented and evaluated, and how they use information in creating their own stories about climate change. We argue that information accessed from the Fox Network and right-wing online sources may deepen confirmation bias but may also provide opportunity for climate communication.

Chapter 6 explores the relationship between climate skeptic identity and feelings about climate change and environmental problems. Anger toward climate scientists, environmentalists, and liberals operates to reify climate change skeptics' positions on certain issues. Yet we also see hints of worry and dread as skeptics fear the negative effects of climate change "if" it is real.

Chapter 7 introduces climate change skepticism as a "continuum" and examines the different ways a skeptic identity manifests among those who actively deny climate science (deniers) and those who are not sure if climate change is occurring and/or if humans are causing it (doubters). The magnitude of a skeptic identity waxes and wanes in accordance with the strength of commitment to rejecting climate science.

Chapter 8 demonstrates the findings of our second round of interviews. When confronted with politically charged issues people tend toward what scientists call confirmation bias, or the tendency to ignore facts that counter one's beliefs. This chapter explores the factors needed to overcome this tendency by sharing the stories of people who have changed their minds about climate change. We explore how identities and ideas shift through processing new information and/or reconciling cognitive dissonance.

We end by exploring the "so what" questions and presenting the implications of our research on climate change communications and the formation of policies on climate change. Though we touch on some of these implications and recommendations in earlier chapters, the conclusion packages a series of recommendations about how to talk with climate change skeptics individually and communicate climate change through more formal venues. We also present best approaches to advancing climate policy and suggest a series of policies that would both mitigate climate change and find support among skeptics. We also examine the implications of identity in the continued ideological fissures in US society and offer ideas for change.

one

SKEPTICISM AS A STIGMATIZED IDENTITY

Nick has spent the entire fifty years of his life in southern Idaho. His parents, both still alive, also have lifelong connections to the region. Though his roots in southern Idaho run deep, of his birthplace Nick says he doesn't "like Jamestown that much. It is an okay place. It has its problems." Reflecting on his time there as a child, he remains similarly stoic: "It is just a normal place. It is just really not anything too special. It's just a place." His relationship with his family is strained, a situation he sums up by saying, "I don't have a great relationship with my mother."

Nonplussed with his hometown and his family, Nick spends his interview talking at length about the ostracism he felt as a child. One example, likely salient for him as he expounded on it without being prompted by us, occurred merely as a result of him supporting the "wrong" candidate during a mock presidential election at school. In fact, it is his earliest political memory. Nick recalls:

> I've always been a news junkie. Actually, I remember Reagan doing his little "Star Wars" speech, and so I have been interested in politics since a fairly young age. The earliest clear memory of participation I think is . . . that election war of [George H. W.] Bush versus Michael

Dukakis. That was embarrassing for me, because I was the only kid in the class who voted for Michael Dukakis . . . I don't know either of these two people and so I voted purely off of the name. And then the whole class tried to hunt down who voted otherwise, which in my case is really bad because my family's trait is big ears and they turn bright red when I am embarrassed. So they caught me real quick. I used the logic against them that they only voted for Bush because that was who their parents voted for, and I voted for Dukakis because his name was Michael, that was it.

Nick's interest in politics, however, stems from a more tragic event. When he was a child, Nick's brother died during an accident that was caused, in his view, by "the underfunding of the public schools, overcrowding of the buses, and everything else." Due to this event, Nick is both cynical and nonpartisan in his political beliefs: "I don't like anyone. Because I know that basically everyone was at fault." He felt like some efforts were made to fix the problems at the root of his brother's death, but then the Republican governor cut school funding again. Idaho Democrats, Nick argues, are "weak" and focused too much on obtaining national office.

We share this information about Nick because his story highlights an important, albeit unexpected, theme that emerges from our interviews with climate change skeptics. By and large the skeptics in our sample organically express a shared sense of marginalization or ostracism from some element of broader society. In Nick's particular case, he conveys a disconnect from his hometown, his family, and politics. He also voices that he felt ostracized as a child for his political beliefs.

This sense of exclusion is at the core of the climate change skeptic identity. Our participants express a shared perception of being discounted in some form in their early lives—a marginalization they perceive as continuing into the present. As we will show, skeptics feel that they are unfairly judged by others (the out-groups) including scientists, liberals, environmentalists, and others, and they argue that they simply want to determine the unadulterated facts regarding climate change. To examine the self-identity of climate change skeptics, we must first understand the basic concepts central to social identity theory and apply those concepts to the case of climate change skepticism. This allows us to analyze the formation of skeptics' self-perception as a collective of innocent knowledge-seekers unfairly excluded from conversations on climate change.

SKEPTICISM AS A STIGMATIZED SOCIAL IDENTITY

In social psychology one of the dominant schools of thought explaining identity is social identity theory.[1] This theory considers identity as a social construction people use to define themselves as members of a group (self-categorization) along with how acquired identities, or perceived group memberships, foster certain intergroup relationships.[2] The development of a sense of self, or self-categorization, occurs through the process of emphasizing the "attitudes, beliefs and values, affective reactions, behavioral norms, styles of speech, and other properties" of other members of the in-group. A given social identity sets a series of acceptable norms, behaviors, and beliefs for adherents. These can be positive or prosocial, such as actions that care for group members, or negative, such as antagonism toward members of perceived out-groups. Social groups also determine appropriate emotional responses to events and stimuli (anger vs. sadness, for example) and behaviors accepted in moments of intergroup tension (online trolling, for example).

Social identities always exist in contrast to out-groups. Thus, to be a climate change skeptic, this identity must be juxtaposed against those who accept climate science. Identification with a group leads people to what researchers call "in-group bias"—the feeling that their group is "better" than another—and a sense of self-worth based on one's part in a group.[3] When tensions arise between groups—in this case between those who reject versus those who accept climate science—hostility between group members often results.

Extending the notion that identification with a group leads to in-group bias and out-group hostility, sociologists Jan Stets and Peter Burke argue that those who accept a group label typically work to distinguish themselves from the out-group.[4] They show more attraction toward the in-group and identify more strongly with the in-group's culture. Holding a group identity can even impact things one might not expect, such as one's perception of disgust, one's food preferences, or one's interpretation of events.[5]

Existing scholarship reveals that environmental beliefs are what psychologist Craig McGarty and his research team call "opinion-based" social identities, founded upon shared opinions.[6] This work is supported by psychologist Ana-Marie Bliuc and her team, who find evidence of group identity among both climate change skeptics and climate change believers: each is rooted in their attitudes toward climate change.[7] These group identities are particularly ripe for in-group bias and intergroup hostility because they are often defined in opposition to one another (skeptics vs. nonskeptics). Under certain conditions, such as perceptions of threat from the out-group, this in-group favoritism may devolve into conflict as well as hostile beliefs and behaviors directed toward the out-group, including distrust, prejudice, and discrimination.[8]

Taking this analysis one step further, we present climate change skepticism as a stigmatized opinion-based social identity. In his 1963 book, sociologist Erving Goffman explored the topic of social stigmas as they relate to identity and explains how stigmas operate:

> While the stranger is present before us, evidence can arise of his possessing an attribute that makes him different from others in the category of persons available for him to be, and of a less desirable kind. . . . He is thus reduced in our minds from a whole and usual person to a tainted, discounted one. Such an attribute is a stigma, especially when its discrediting effect is very extensive. . . . The term stigma, then, will be used to refer to an attribute that is deeply discrediting.[9]

Goffman presents stigmas as related to physical distinctions, questions of morality, and group affiliations (religion, race, etc.).[10] He also distinguishes between those that are obvious/visible (discredited) and those that can be disguised (discreditable). Goffman's successors, criminal justice professor Mark Stafford and collaborator Richard Scott, extend Goffman's definition to define the stigma of invisible markers or group membership: "The [stigmatized] characteristic may involve what people do (or have done), what they believe, or who they are (owing to physical or social characteristics)."[11]

Climate change skepticism can be seen as what Goffman viewed as a discreditable stigma based on perceptions of moral character or a "counter-normative identity," a stigmatized identity wherein they "hav[e] characteristic[s] deemed contrary to the shared beliefs and expectations in a given situation held by members of a social unit."[12] Consider, for example, the numerical minority position that skepticism reflects in society overall. Even though the United States has the highest percentage of residents who do not believe that global warming is happening (at 14 percent, while 72 percent do believe it is happening and another 14 percent are not sure/do not know/refused to answer), skeptics are a counter-normative minority when considering the broader American public.[13] Indeed, psychologists Matthew Hornsey and Kelly Fielding lay claim to the assertion that climate change skeptics are "subject to mainstream ridicule" and that this identity is both "small and stigmatized."[14] This stigmatization, Hornsey and Fielding argue, serves to reinforce the sense of connection people feel with the identity and can positively impact their self-esteem.[15]

Our interviews with self-identified skeptics reflect exactly what Hornsey and Fielding describe. Participants expressed a sense of victimization due to their perceived marginalization; skeptics believe that they are excluded, dismissed, and degraded by the mainstream society and by those who believe in climate change, but especially by climate scientists, environmentalists, and liberals.[16] Skeptics' perceptions that they are derogated by people who accept climate change is supported in the comments made on internet news stories. In her dissertation, Angela Brodsky examines the comments made on Yahoo! Forums with regards to climate change skeptics. Brodsky finds that those who believe in climate change perceive skeptics as "ignorant, uneducated, and scientifically illiterate."[17] This perception and experience of rejection and exclusion appears to elevate skeptics' sense of connection to other skeptics and the pride they take in the identity itself.

In the following sections we demonstrate the salience of early exclusion in the minds of the climate change skeptics in our study. We then trace how this sense of ostracism remains relevant in the present and is intimately connected to the identity of climate change skeptic, thus creating a perception that skeptics are a stigmatized social group. We conclude by exploring how skeptics reframe themselves in light of this stigmatization and claim that scientists are exclusive and corrupt while skeptics themselves are innocently seeking unbiased knowledge in the face of overwhelming societal bias against them.

LONG-TERM EXPERIENCE AS OUTSIDERS

In recalling their childhoods, participants shared experiences of feeling excluded. For many, this was a sense of being ostracized as political outsiders; for others, as religious outsiders or as residents in isolated rural communities. Regardless of the cause of the ostracism, these experiences emerge as essentially relevant, as participants draw connections between their early marginalization and their self-perceptions as stigmatized today.

Take Nick, discussed at the start of this chapter, for example. Nick opens his interview by telling us about his sense of disconnect from his family and community, as well as an early memory of political exclusion. His stories were in response to the questions "Did you like growing up there?" and "At what age do you remember becoming aware of politics?," which indicates the level of salience these memories and feelings hold for him.

Douglas, in the introduction, grew up in Appalachia. The son of two schoolteachers, he recalls having a strong social community as a child, with weekly church services and social events, including Boy Scouts, soccer, and other activities. He has two brothers very close to him in age, so he always had two playmates at home. Though he felt connected to his family as a youth, that sense of acceptance did not carry over at school. When asked if he enjoyed school, he recalls, "I was pretty bookish growing up, kinda socially awkward." He performed well in school, but he struggled with creating and maintaining friendships with his peers. His bookishness served him well, as he ultimately followed in his parents' footsteps and currently works as a biology teacher.

The primary feeling of exclusion Douglas feels is grounded in his political ideology. He feels he never truly found a political home or community. Douglas remembers his first awareness of politics came during the 1984 Reagan-Mondale election, when a friend asked him to help remove bumper stickers from people's cars. While this first memory was of a time shared with one of his peers, his sense of not fitting in within the political system became salient as he grew and his ideas about the world solidified. Douglas told us that he always cared deeply about sustainability but that he has struggled to find alignment with either of the two main political parties in the United States. Douglas spent time exploring the Green Party, the Republican Party, and others. He now identifies as a "Christian, increasingly apolitical, socially conservative, environmentalist," but still wrestles with finding a place where he is ideologically accepted. His environmental views are often not shared by fellow conservatives nor are his thoughts on climate change welcomed by his colleagues in environmental action. Further, his family rejects his recently found conservativism: "My brother has recently become more of, say, a liberal . . . he would try to pick fights with me because he knew I was increasingly conservative." Like his brother, Douglas's parents have also become more politically progressive over time, particularly recently, in response to former President Trump's anti-immigrant rhetoric and policy initiatives.[18]

While Douglas primarily reflects on feelings of political exclusion, other people we talked to express feelings of being outsiders because of their religious views. Jack was born and raised in Idaho. Unlike most people in the state, Jack identifies as a Democrat. In his words, this is because "I have never, ever been rich . . . I earned wages and have never made good money." His financial struggles are partially connected to what he perceives as religious discrimination. When asked about his political ideology, Jack shares an experience of being fired because, at least in his perception, he did not share the religious convictions of his supervisor:

> A new person bought the company that I was working for. I had been working there for five or six years and had accomplished a little bit of position in the company, and the first thing that he did was clear out a lot of people for two reasons I conjecture. The first being that it was a time of tough employment, we were just going into the Bush depression, and so finding replacement drivers was not a problem and paying them a lot less than what I was making was certainly something to encourage him to replace a lot of us. . . . However, there were other people . . . who had been given jobs . . . they belonged to the same religion as the owner, but I did not, so . . .

Jack experienced what he sensed was discrimination based on his religion. He felt that he was fired because he did not share a religious affiliation with his boss.

Other participants also perceived exclusion based on religious ideologies. Savannah, for instance, grew up in northern Utah and southern Idaho. Even in this predominantly Latter-day Saint (LDS) region, Savannah recalls the way her religion set her apart from her community. Her fundamentalist LDS beliefs were "not the norm." She recalls that her "religion wasn't widely accepted, which made us kind of alone, I think would be a better term. We were alone. Our opinions, our beliefs, everything was centered around just family, which made it hard to connect with other people." To exacerbate this feeling of exclusion, Savannah's family also lived outside of town, and she and her siblings were homeschooled, so she had very little contact with anyone besides her family. In the few years Savannah tried attending public school, she felt she was "socially awkward" and couldn't "quite figure out how to behave or even understand people of my own age." As a result, she had a hard time "being with other people" and "didn't participate in a lot of stuff." Nonetheless, Savannah loved learning and the curricular side of schooling. Things changed a bit when she got married (at age sixteen) and was able to more fully explore her personal beliefs about politics. Now, like Douglas, she feels she doesn't have a true political home. She finds agreement with conservative positions on certain topics such as abortion, but agrees with liberal positions on others, such as conservation and gender norms.

These experiences—professed exclusion in childhood based on political views, religious beliefs, geography, or even a sense of being "socially awkward"—provide early memories of times when people who now identify as climate change skeptics have felt marginalized in some way. Given the salience and consistency with which these memories emerged in our interviews without any direct prompting, it is clear that a shared sense of marginality is a key aspect of identity as a skeptic.

PERCEIVED PERSECUTION

The climate change skeptic identity is rooted in a sense of perceived marginality. In interviews, participants organically relayed salient memories of exclusion from the past and continuing into the present. They then focused on the ways that they have felt dismissed or excluded by others because of their skeptic identity and associated beliefs.

Tyler, though originally from Colorado, spent significant time as a child in Idaho at his family cabin. He became aware of politics at a young age, often working with his mom on different political lobbying efforts. Tyler is what climatologist Stefan Rahmstorf calls an "attribution skeptic."[19] He believes that the climate is changing but sees this change as a natural part of the earth's cycles and not a result of human activity. He bases this perception on what he sees as a lack of reliable data used by climate scientists. But Tyler's real passion, and what he repeatedly returns to in his interview, is how he feels climate change skeptics are treated in today's society. When asked about his beliefs on climate change, Tyler launched into a monologue about how people with questions about the veracity of anthropogenic climate change are treated unfairly. He argues, "People simply will attack you for the notion that you've actually questioned what could be considered the status quo around climate change or anything like that." He continues: "Anytime you question any one single part of what is going on with climate change, you're considered persona non grata and . . . your entire voice and reason is rejected." Like many people in our sample, he particularly calls out celebrities who he views as unfairly dismissive of skeptics. For example, Tyler implicates Bill Nye, "the Science Guy," claiming that his position on climate change is, "I'm Bill Nye, if you don't listen to what I have to say, you're an idiot and no one's going to listen to you."

Unlike Tyler, James is not convinced that Earth's climate *is* changing. An evangelical pastor who resides in rural Idaho, James believes that the planet is a gift from God and that He will continue to protect it, so there is no need to worry about climate change. Despite differing in the nature of their skepticism, Tyler and James both share frustration with what they see as the immediate dismissal of critics of climate science and the lack of civil discourse on the subject. James argues: "On the subject of climate change, when someone stands up at a conference and says, 'I don't believe it, I don't believe it.' The person who does believe it, doesn't stand up right after and say, 'It's really important and healthy that you be able to say that.' No, what happens isn't that. It's an outrage. It's a scandal." He believes that "highly qualified skeptics . . . have been kind of blackballed, blacklisted."

In their aforementioned work, Hornsey and Fielding explain that the critique leveled by mainstream groups against stigmatized groups can be used by the smaller collectives as further evidence that the mainstream is "deluded and/or malicious" and reinforce trust in the in-group (in this case, climate change skeptics).[20] This is evidenced in the way that skeptics talk about the personal danger or persecution they face for their beliefs. Douglas suggests that skeptics are "demoniz[ed]" for their perspectives and compares this martyrdom to that experienced by Galileo, arguing that people have been "jailed for their beliefs." James posits that the term denier, when applied to a climate skeptic, is used to intentionally align skeptics with Holocaust deniers. He says:

> Climate change skeptics became deniers in the public debate. So, I'm a climate change denier now. Well, what is that supposed to evoke? That is supposed to make you think of Holocaust denier. The word "denier" is someone who is a crank. Sort of out on the outer edges of polite discourse. Holocaust deniers are shunned; we don't even engage with them; we don't have to engage with them. And you guys are climate change deniers, or deniers, for short, which tells me that you don't want to debate, you don't want to discuss it, you want to shut the conversation down.

At the most extreme edges, skeptics argue that they are persecuted to such an extent that they risk losing their jobs or even their lives for expressing their beliefs. For instance, David—James's brother and a biology faculty member at an unaccredited evangelical college—alleges that skeptics "have to worry about their personal safety" and "have to start fearing for their job, fearing for their personal well-being." He then compares climate scientists to terrorists blinded by their passion: "Even if it is all based on a false premise, you can drum up a lot of fervor and it's like you see in the Middle East. You get jihadists. It can get scary. That's what bothers me." Here he alludes to the possibility of skeptics being physically harmed because of their beliefs.

Nick, introduced earlier, is more direct. He suggests that President Obama "start[ed] ratcheting up to jailing dissent to things like that," which he compares to both the French Revolution and a "reign of terror." He claims that, under the Obama administration, "stuff started ratcheting up to jailing people [over] dissent to just killing them and executing them." He continues by expressing fear of the same passion that David critiques:

> Even if you believe climate change is real, that viewpoint is toxic . . . That's mass hysteria levels, and we always screw up when we get there. Even if it is the right thing to do—or kind of—we always tend to go to some extreme. It's one of those—you don't want to get a large group of people into that mode because then they will just start pulling people from their homes and killing them because of any nonsense reason. It's not acceptable.

These claims indicate that skeptics believe they are targeted by the mainstream society through threats of violence, incarceration, and job loss. In truth, we found no evidence to suggest institutional or widespread threats or attacks being made against people who reject climate science. Instead, evidence points in the opposite direction, to what climatologist Michael Mann calls "modern-day McCarthyism."[21] In fact, the Federal Bureau of Investigation reports a measurable increase in "threatening communications to climate scientists."[22] These include emails that threaten violence, encourage suicide, and in some online spaces, express racist propaganda.[23]

The exaggerated sense of persecution skeptics feel serves to reinforce their in-group identity and their perspective of climate scientists and even the general public as evil. This reinforcement in turn seems to bolster skeptics' sense that they are honorable victims. Here again, work by Hornsey and Fielding sheds light on this phenomenon: "In these circumstances, groups can 'bunker in,' forced into insularity by the ridicule of the mainstream, but emboldened and sustained by that very insularity."[24]

SEEKING THE TRUTH

Climate change skeptics construct an identity of marginalization while simultaneously contending that they are simply seeking the truth regarding climate change. They lament what they perceive as a vilified exclusion from public discourse about climate change and argue that they simply want the data. Indeed, when we asked how they self-identified, skeptics in our survey used language such as "truth finder/seeker," "fact finder/

seeker," "logical thinker," "critical thinker," "investigator," "evidence-based," "questioner," "proof seeker," and "learner."

Consider Sam's comments. Sam grew up in the Midwest and moved to Idaho as an adult. He currently works as a Catholic missionary. He first expresses frustration with exclusion and then makes a claim to moral motivations: "Th[e] people who adhere to climate change [argue] that if you have any skepticism you're a climate change denier. You're antiscience. You're all these different things and it's kind of like, well, I want to actually figure this out, I don't want to just be told." This suggests that skeptics believe themselves to hold high moral regard for the truth but distrust the people who produce knowledge. David, the biology faculty member at a Christian college in northern Idaho who rejects anthropogenic climate change, also defines himself in these terms: "I've always been a critical thinker" who wants to "objectively" look at the data.

Like Sam, Peter moved to Idaho from the Midwest. As an "impact skeptic" who believes that the earth's climate is changing but that it will not have a negative impact on humanity, Peter expresses similar views.[25] When we ask Peter if there is any information that might change his mind about climate change, he responds that he would be more willing to accept climate science "if there's more healthy debate out there." Instead, he laments the tendency of climate scientists and others who accept climate science to "just label people instantly as 'deniers,'" which eliminates the possibility of "any healthy debate, or conversation around it." Peter and other skeptics specifically reject the extrapolation of data in predictive models. "I'm always skeptical [when] people predict the end of the world." He argues that this is akin to "crazy religious leaders that kill all their followers." Ultimately Peter reiterates that he would accept climate science data in the presence of a "more healthy debate, instead of just calling everyone a denier if they make any questions."

While skeptics express a variety of concerns regarding their safety and inclusion in public discourse, central to their identity construction is an alignment with being data-driven. Sam argues that he is dedicated to investigating climate change and will do so relentlessly and with integrity: "I want to keep looking into this, I want to keep seeing the information that's out there, I want to keep hearing from atmospheric scientists, and climate scientists, and be informed on the issue." James, too, just wants to see the data: "Basically [it's the] data. And if it's data that submits to cross-examination. Everybody claims data, but [I want] data that's not coy, that's not shy, that's willing to have a debate." Douglas seeks better understanding of climate modeling: "I'd love to better understand and I hope to spend more time doing this. The models, especially as they're updating continually, [it's] hard to keep track of what the current projections are."

At the crux of skeptics' identity construction is the perception that they are ostracized and excluded from the public discourse on climate science; they believe that they are a marginalized group facing persecution. Skeptics also contend they face job loss, imprisonment, or even death because of mainstream persecution for their perspectives on climate change. They see themselves as victims of the dominant society—as truth-seekers in a society that works to exclude them and prevent them from uncovering reality. In the creation of a social identity of ostracized truth-seekers, they reinforce their own identity but meanwhile bolster their perspective that climate scientists are unethical, greedy, and even evil (a topic we turn to in the next chapter).

It is worth mentioning that one's sense of self and others need not be objectively true for it to impact one's attitudes and behaviors as well as one's relationships with others and with information. Rather, this is precisely the power of identity and the power of social stories more broadly: it helps convince people of a particular reality, which in turn serves as the foundation for their interpretation of life more broadly.

THE IMPLICATIONS OF A STIGMATIZED IDENTITY

In an interview with popular podcast host Shankar Vedantam, psychologist Jay Van Bavel examines the influence of identity on connection with others, with one's preferences, and with one's interpretation of events. Van Bavel explains that "one of the most powerful ways to trigger an identity is to be a minority in a situation." As a numerical fraction of the larger population, skeptics have formed a shared identity based on their perceived marginality.

More specifically, climate change skepticism is an opinion-based stigmatized social identity that emerges in opposition to a perceived out-group: climate scientists. Skeptics construct themselves as people who have been excluded in childhood, who continue to be dismissed and vilified for their beliefs in the present, and who simply want to find accurate data without the "hype." Through their vilification of climate scientists, they construct themselves as noble truth-seekers, a label that becomes self-affirming in the event that they encounter resistance to their beliefs. It serves to strengthen their self-identity as skeptics in any instance in which they encounter information or resistance to their belief structures. This reinforcement makes the process of communicating facts related to climate change more difficult by increasing the tendency toward confirmation bias (i.e., the preference for messages that align with preexisting attitudes over those messages that challenge them).[26]

Accepting an identity affects one's behavior, values, and beliefs. It dictates what is doable and sayable and even feelable in a given context. Once a person identifies with a group, that person behaves in ways that reinforces a specific sense of self.[27] Their actions serve not only to reify these self-beliefs but also to signal to others how an individual should be seen.[28] These behaviors do not operate in isolation but are part of a broader culture or structure.[29]

The possibility of climate change skepticism being a social identity has significant implications for communications about the climate. Because of the deep connection one feels to the created self-perception as a skeptic, simply presenting facts or data about climate change is unlikely to affect a skeptic's position on the issue, despite what skeptics say about simply "wanting the facts." Instead, such challenges may be perceived as further evidence of the deceptive nature of climate scientists and the tendency for the public to be easily duped, thus reinforcing the noble identity of the skeptic. Confirmation bias, then, amplifies the very polarization on the topic of climate change already operating.[30]

In their work on polarization, communication scholars Shanto Iyengar, Gaurav Soon, and Yphtach Lelkes find that affect, or emotions, contribute to division more than ideas. They connect this tendency to a "primordial sense of partisan identity" learned by adherents, which then shapes all encounters with perceived others. Iyengar and the team contend that these affective forces are fed by media and political campaigns that add fuel to such animosities. They further contend that affective divisions prevent people from seeing the good in "others" or recognizing positive policy implementation, instead attributing all decisions to their duplicitous nature. This dynamic undermines the health of political debate and even democracy itself.[31]

While our primary project is to examine and understand the nature of climate change skepticism in the United States, we use that understanding and tension to assess the possibilities for shifts in identities, attitudes, and collective action on the climate. Toward that goal, we turn to an examination of skeptic perceptions of climate science and scientists.

two

(DIS)TRUST OF SCIENCE

BILL, A LIBERTARIAN WITH A COLLEGE DEGREE, grew up in Spokane, Washington. He and his wife moved to Idaho about twenty years ago and now live, in Bill's words, "about as far out in the country as you can get and still have high-speed internet." First employed at a local newspaper, Bill later obtained a college degree and launched a second career selling real estate. By the time we interviewed him in 2018, Bill had left his real estate company and was working at a local big-box store.

Bill has a slew of reservations about climate change, many of them related to his distrust of climate science and climate scientists. When we asked Bill about these misgivings, he talked about "leaked emails that indicated that they [climate scientists] had deliberately ignored some of the results that would have disproved their theories." Referring to the 2009 "Climategate" scandal in which an email hack led to false claims by climate change skeptics that scientists manipulated climate data, Bill further elaborated:

> Now some of these climatologists argued about how the emails were obtained but they did not dispute the results of these emails, what the emails contained, and so that got me thinking. You know, if scientists are deliberately skewing data to indicate one result or another, can you really trust their results? And, so, it was back then that I was thinking maybe this climate change stuff isn't caused by humans, maybe somebody is or some group is deliberately manipulating

> the data in order to produce the results they want, and since then I've heard of a few other articles that basically help to add to that theory . . .[1]

Bill's distrust of climatologists extends beyond a belief that climate data are being fabricated to support the dominant scientific view on climate change. He is also concerned that incentive structures in science, particularly the funding of scientific research, and the politicization of scientific findings influence climate research, arguing, "I would say it's a matter of following the money. Who's paying for these studies? . . . Some of these climatologists who were saying that manmade climate change is a definite fact, some of them are being paid by government studies, and governments, of course, use this data to go ahead and pass legislation to do whatever they want." Bill further contends that "some of these scientists and their studies have been paid by some multinational oil conglomerate. I mean, unfortunately, the study of climatology has been so politicized that it's hard to tell who to trust anymore, really."

Bill compares climatology to astronomy, which he sees as a more objective field of scientific inquiry, free of financial and political influences. "There's really not a whole lot of political motivation to discover that there's a planet orbiting the sun that seems to be made out of pure diamond, so." In contrast, Bill concludes, "in regards to climate scientists, I'm not sure if anything can be trusted in that field anymore."

Bill's distrust of climatology and climate scientists is not unique; it mirrors the thoughts and concerns of many climate change skeptics. In our interviews and surveys with skeptics we find that many express similar concerns: climate change may not exist or it may simply be a natural cycle of warming. These concerns are rooted in skeptics' attitudes toward climate science and scientists. In fact, our research reveals that doubts about climate science and scientists drive skeptics to construct their identities in opposition to climate scientists. Distrust compels skeptics to view scientists as an untrustworthy, uncredible, and exclusive out-group; an out-group they blame for constituting one axis of their perceived marginalization in the climate discourse.

In chapter 1 we explored how skeptics see themselves—the creation of their self-identity as innocent truth-seekers persecuted by broader society, but especially by climate scientists. In this chapter we scrutinize climate change skeptics' perceptions of climate science and scientists. We begin with a discussion of scientific trust: the factors that contribute to trust in science and how that trust shapes public opinions of controversial science topics, including climate change. We then show how climate change skeptics construct meaning around climate scientists as an out-group, and the us-versus-them relationship that results. We conclude by discussing the implications of this distrust for climate change communication and policy.

TRUST IN SCIENCE

People in the United States generally have high confidence in the scientific community. In fact, the scientific community ranks second as the institution in which most Americans hold confidence, just behind the US military.[2] According to the National Science Foundation's Science and Engineering Indicators report of 2020, many Americans (44 percent in 2018) continue to state that they have a "great deal of confidence" in the scientific community, a percentage that has remained stable since 1973, when the data was first collected.

Even in the wake of public debates on science surrounding the COVID-19 vaccines, 64 percent of Americans still report significant levels of trust in science, according to a 2021 Gallup poll.[3] Most Americans also find social benefits to science, believing that science generates opportunities for the next generation (92 percent in 2018) and that government should continue to fund basic scientific research (84 percent in 2018). However, there are clear political divisions as well. While in 1975 Republicans reported greater trust in science than Democrats, today only 45 percent of Republicans report that they trust science (down from 72 percent in 1975) compared to 79 percent of Democrats (up from 69 percent in 1975).[4] Indeed, emergent scholarship suggests that trust in science may be the leading factor in the partisan divide on climate change.[5]

While overall trends are encouraging from the perspective of public investment in scientific research and social acceptance of new scientific innovations, over the past decades several scientific topics have also become controversial in the public's eye.[6] Increasingly contentious topics include vaccines, genetically modified food (GMOs), and climate change.[7] Despite such concerns, all three possess what is called "scientific consensus," or the near unanimity in agreement among scientists on their efficacy and safety (vaccines and GMOs) or the existence of the phenomenon (climate change).[8]

While public debate continues, there is overwhelming scientific consensus regarding the anthropogenic origins of climate change. In fact, by 2019 climate scientists worldwide had reached near unanimity in their acceptance of the realities and human causes of climate change. Nevertheless, a significant fraction of the American public remains unconvinced. For instance, according to researchers at the Yale Program on Climate Change Communication, in 2021 only 57 percent of Americans thought scientists agreed on this issue.[9] Among our survey respondents, the percentage was even lower: only 48.6 percent of respondents agree that most climate scientists do believe in anthropogenic climate change, and 38.6 percent agree that "there is no scientific consensus that human-caused climate change is happening." The difference in these percentages is likely explained by the fact that our study participants are self-identified skeptics only, while researchers at the Yale Program, including their well-known "Six Americas" study, surveys the general public.[10]

Regardless of whether or not they believe that scientists agree on climate change, the skeptics in our sample expressed higher levels of distrust of scientists. Only 35 percent of our survey respondents thought scientists warrant any level of trust at all. Such distrust is problematic because trust of and belief in the credibility of the scientific community are key factors that shape public attitudes.[11] For complex topics such as climate change, even direct experience can be significantly influenced by the credibility of scientists who conduct and communicate their research and the media that disseminate their findings to the wider public.[12] Particularly in situations where people are presented with new or conflicting information, people tend to rely on their trust of or the credibility of the messenger, since both serve as a mental shortcut when determining the validity of scientific claims.[13] Trust provides the avenue by which people map out appropriate courses of action.[14]

WHAT SHAPES TRUST IN SCIENCE

Factors associated with trust in science include a combination of social factors and certain value orientations. For example, in his work sociologist Gordon Gauchat shows that sociodemographic variables such as political orientation, religiosity, education, and degree of scientific literacy is correlated with trust, wherein Democrats, nonreligious people, and those with higher levels of education and greater scientific literacy are more likely to trust science than their respective counterparts.[15] Exposure to information also impacts one's confidence in science. For instance, in their work on media, researchers John Besley and James Shanahan find that stories told through media (specifically television) drive public confidence in science and people with greater exposure to science reporting develop greater trust in science.[16]

Trust in science, however, depends on the topic in question. Democrats, for instance, are more skeptical of genetically modified food, while Republicans have less confidence in human-caused climate change.[17] An individual's trust in science is also influenced by their perception of whether science policy is shaped by scientists or by politicians.[18] For example, people may hold different levels of trust in production scientists (whose research leads to new products and technologies) as opposed to impact scientists (whose research identifies harmful effects of new technologies and products on the natural environment or human health).[19]

In their research on scientific trust, sociologist Lawrence Hamilton, environmental scientist Joel Hartter, and sociologist Kei Salto examine the factors that shape confidence in scientists being a reliable source of information about climate change (among other contested scientific topics). They find that political ideology and education matter, but inconsistently. Among Democrats, higher levels of education correlate with greater trust in scientists as an information source.[20] Among Republicans and Tea Party members, education has no significant effect on scientific trust. Typically, there tends to be more concern about climate change among people who trust scientists as information sources as opposed to those who do not.[21]

We find that climate change skeptics who are politically conservative tend to have higher levels of distrust in climate science and scientists compared to their more liberal counterparts.[22] This same research demonstrates that skeptics who are men and those who are more religious have lower levels of trust compared to their respective counterparts. Crucially, then, the factors that shape distrust in climate science and scientists (e.g., political orientation, gender, religion) are similar to the factors that lead to greater climate change skepticism. This indicates that distrust likely constitutes an important element of skepticism among self-declared climate change skeptics in the United States.

DISTRUST AND AN OPPOSITIONAL IDENTITY

For skeptics, the threats caused by climate change are incongruent with their beliefs about climate science. Skeptics see both the narrative of climate change itself and those who advance it (climate scientists) as threatening and even deceitful. Such feelings exacerbate distrust toward scientists who are positioned as a clear out-group.[23] Suspicion is reinforced by right-wing politicians, contrarian scientists, and the conservative media outlets that promote messages in line with skeptic thinking on climate change and science (see chapter 5).[24] From a skeptic perspective, climate scientists are dishonest: they are pressured to advance the mainstream position on climate change by pervasive incentive structures in scientific research, they use unreliable data and methods to draw their conclusions, and they exclude from public debate anyone who disagrees with them.

Belief That Scientists Are Pressured to Toe the Line

The skeptics we interviewed argue that climate scientists are untrustworthy because they are influenced by what they perceive as widespread incentive structures that bias scientists and the work they produce. Of particular concern to skeptics are those who fund climate change research, arguing that institutions that fund research (such as governmental or international organizations) exert undue influence on climate scientists, on their research agendas/outcomes, and on what gets published, both overtly and covertly.

Take David, for example. A politically conservative evangelical Christian, he believes that funding agencies essentially coerce scientists into producing outcomes that align with agency agendas. He describes this concern by saying, "If you are an academic that wants grant money and your livelihood depends on it, your tenure depends on it, you know, lots of things are hanging in the balance." David continues: "It's not driven by this objective desire for truth, it's driven by—this is what scares me as a scientist—agenda-driven science scares the bejeebers out of me—because people, uh, just go with where the money is."

David is a particularly illustrative participant. He has a PhD in environmental science and works as a science professor at a small religious college. Despite his scientific background, David distrusts the process of science and science funding. In fact, he reflects what social scientists Caitlin Drummond and Baruch Fischhoff find in their work examining the correlation between education, scientific literacy, political views, religion and polarizing scientific topics. Drummond and Fischhoff find high levels of political polarization among people with higher levels of scientific education and literacy. For political conservatives, greater scientific education and literacy is associated with lower acceptance of climate change, whereas for progressives, scientific knowledge increases belief in climate science.[25]

The narrative that scientists are unduly affected by funding is something shared by skeptics across educational backgrounds. For example, Douglas (introduced in chapter 1), who has a college degree and works as a biology teacher at a regional charter school, shares David's concern about funding, saying, "Well, if someone is funding you who leans toward one way, you're going to be . . . you're going to want to support more of what they think so that they don't cut your funding."

In fact, nearly everyone we interviewed expressed this concern regarding research funding, though some responses varied in the degree to which they thought such coercion was exerted overtly or covertly. Mark, a white atheist college graduate in his sixties who identifies as libertarian, believes that scientists deliberately engage in unethical behavior to fund their research: "If a professor wants to get a grant to do a study, he's got to, depending on who's funding it, they're gonna want 'this is the result we want, we'll give you the money if you come up with this answer.'" Other respondents believe that the influence of funding is less apparent. Zed's perspective is that scientists are "generally trustworthy." A participant from Boise Valley, Zed believes that at times scientists "get pushed by the funding," which "forces them to leave things out that may make a difference." He believes that funding agencies "lay conditions down," such as requiring scientists to not use funds to investigate certain related topics, which makes it "hard for scientists to work."

While many skeptics believe that the influence of funding on science is pervasive and nefarious, there is little evidence that funding exerts significant and consistent impact.[26] There are, of course, exceptions. In a few rare cases, university partnerships with the pharmaceutical industry, for instance, has led to misconduct, such as the failure to disclose conflicts of interest.[27] Tellingly, skeptics generalize from these anomalous cases to suggest that most or all scientists are untrustworthy.

Ron shows how this logic operates. Ron grew up in California and currently resides in southern Idaho. During our interview he expressed concern about the trustworthiness of scientists by explaining, "I mean, everybody has a price. I mean, everybody can be bought, everybody can be scared into being quiet or talking about things so, I mean, how are you supposed to know in this day and age?" As evidence for this position, Ron cited malpractices in the pharmaceutical industry and among medical professionals, his reason for believing that "everybody" can be influenced by money, further stating that "all the doctors, everything is this revolving chain of money." Ron continues to hold on to this belief, although actual cases of misconduct, beyond certain high-profile cases that have garnered media attention, remain exceedingly rare.[28]

Going beyond the unethical influence of funding, skeptics also believe that scientific journals that publish research on climate change are biased. According to skeptics, these journals prefer articles with positive (that is, statistically significant) and sensational results, which ultimately promotes a consensus view on climate change trends, causes, and effects. As a result, in addition to viewing scientists as untrustworthy, skeptics also lack confidence in scientific literature.

Nick, the conservative Democrat from Boise, elaborates this concern by discussing what he calls "new research studies." According to Nick it is hard to publish "null results." To justify funding, scientists must publish only projects that result in evidence that support their hypotheses, which they then sensationalize. He explains: "You said you found nothing . . . no one publishes studies like that. And so we have this weird bias in science where the only things that are published are the things that say things that are sensational." Nick presents renowned climate scientist Michael Mann (a Pennsylvania State University professor) as someone who has similarly sensationalized his findings but is now attempting to "taper down on some claims that he's created because he realizes that some of the extreme things are actually doing damage to his own argument. Because they're not happening."[29] Mann rose to prominence in 1999 for documenting the sharp increase in global temperatures since the Industrial Age, in what has become known as the "hockey-stick graph." Mann is a frequent target of climate disinformation campaigns and as a result is widely distrusted among skeptics like Nick.

Nick's concern about publication bias (that studies with positive results are more likely to be published), while inaccurate, is not entirely unfounded. In certain scientific fields, particularly in biomedical sciences, researchers have found a bias that occurs at the time of submission, when authors tend to be less likely to submit articles that report negative results. However, once submitted to journals, no significant differences have been uncovered in publication rates between articles with positive and negative results.[30] Furthermore, in climate science, researchers have found no evidence of a publication bias in terms of underreporting of nonsignificant results.[31]

Along a similar line, James, the evangelical pastor and political libertarian with a college degree, questions the trustworthiness of scientific journals on the basis of the peer review process. James believes that this process is not entirely blind: "It's gotten kind of cozy and everybody knows everybody. So refereed journals aren't refereed as tightly as they used to be." In James's perspective the peer review process also "runs out" scientists whose research and publications do not align with the dominant view in the field. "And, if you say something heretical, basically we have—the scientific establishment has become an orthodoxy and they persecute. If you say something heretical, they run you out. You get denied tenure or you're threatened with denial of tenure."

Concerns about explicit and implicit coercion through funding, unethical publishing practices, and career advancement pressures constitute a broader concern among skeptics: that climatology is overwhelmingly biased toward generating and maintaining a certain consensus view on anthropogenic effects.

Bob grew up and lives in a small college town in northern Idaho. He suggests that this consensus view is so powerful that dissenting scientists are marginalized as "crazy." To Bob the consensus narrative of climate change serves the ultimate goal of granting governments greater control over the public. On climate scientists, he believes that institutional coercion is such that even some scientists who do not agree with the dominant view will still adopt it: "And, I don't know, there's the occasional one or two, might just jump on board because their colleagues just say, like, if you don't believe in this you're crazy, which is the general narrative too." As evidence of this effect, Bob cites a story of a scientist who denied the Big Bang happened and was labeled "crazy" by his colleagues. "And so he's ostracized by the community, by the scientific community, by doubting the general consensus of what you know is just a theory." As a result, the "majority of scientists don't want to doubt what everyone else wants to claim is real."

James, the evangelical pastor with a college degree, agrees: "Others know that it's not true but they don't want to ruin their careers by becoming a denier. If I get tagged as the denier then I might not get tenure, I might not, you know. So, I'll just keep my head down until this thing blows over."

Because skeptics believe that incentive structures in science forces climate scientists to advance a consensus view, skeptics express higher trust in scientists who do not agree with the consensus. Accordingly, "it takes a lot of guts to step outside the herd." James explains:

> I sometimes trust, or I listen to, the minority because they're cutting against the grain. They've got a lot to lose. And it is easy to just go with the flow, if the consensus says this and grant money is being showered upon you because, hey, you're a climate change believer, you know, there's, it's easy to go with the flow. But if you cut against the grain because you have some convictions that it's not a slam dunk, I listen to that.

Skeptics' view that climate scientists use explicit and implicit coercion to maintain a consensus view on anthropogenic climate change does not align with existing research on the topic. For example, in their book *Discerning Experts: The Practices of Scientific Assessment for Environmental Policy*,

Michael Oppenheimer and his colleagues find the exact opposite; that instead of maintaining an alarmist consensus view regarding human causes, scientists err "on the side of least drama" and potentially underestimate the pace and severity of climate change.[32] In other words, the reality might possibly be even more dire than the presently reported scientific consensus.

All in all, the concerns among skeptics that climate scientists are swayed from honest and ethical work due to prevailing funding structures, that publication biases exist, and that career pressures and pressures to advance a common narrative on human-caused climate change are the rule of the day, make them largely distrustful of climate science. In fact, among our survey skeptics, a majority (57.3 percent) agree that climate scientists are influenced by funding. Only 13.5 percent disagree with this statement, and the rest (29.2 percent) state that they neither agree nor disagree. A significant fraction, 46.8 percent, agree that climate scientists experience pressure to "make certain claims in order to get tenure." Nearly 40 percent agree that scientific journals are biased toward publishing studies that show climate change is happening, while another 40 percent take no position on the question. It appears that distrust of climate scientists has become pervasive and is a critical piece in the construction of a skeptic identity; concurrently, skeptics construct scientists as the "other," a clear out-group that is easily swayed by personal and institutional incentives.

Belief that Data and Methods Are Questionable

When co-constructing climate scientists as the "other," skeptics resort to questioning the credibility of scientific practice around climatology. They argue that climate scientists do not have enough data to determine the trends, causes, and effects of climate change and that scientists use inadequate or faulty data, engage in various data manipulations, and sometimes fabricate data outright. Skeptics argue that climate scientists do not accurately follow the standard scientific method and that climate modeling is not a legitimate form of science, but is rather "educated guessing."

Among our survey respondents, 22.5 percent of skeptics agree that "climate scientists aren't doing real science" (37.8 percent remain neutral while 39.7 percent disagree). A comparable fraction, 25.9 percent, agree that "climate modeling isn't science" (43.1 percent neither agree nor disagree with this statement). A significant fraction of skeptics (29.5 percent) also agree that "doing climate science is the same as making educated guesses."

To elaborate some of these perspectives, consider James, who was introduced earlier. James argues that climate scientists have not been measuring global temperatures long enough or accurately enough to make any broad claims about temperature fluctuations. He raises a slew of related questions: "So the first thing I bring up is, okay, how long have we been measuring temperatures accurately? How many years? Better yet, how many months have we been doing this? How long have we had thermometers all around the globe in the same place, in secure places that give us an accurate reading?" Adding his religious perspective to the discussion, James suggests that even taken from a young earth creationist perspective, believing the planet to be just 6,000 years old (as opposed to the scientific estimate of 4.5 billion years), scientists have only consistently and accurately measured temperature data for a fraction of that period. He argues: "Less than a fraction of 1 percent, that's what. And, so, you're going to look at this little fluctuation, the joke's told for a little boy who, a little boy who gets a hammer, everything looks like a nail. And we just got these thermometers put in place and we are hovering over these fluctuations, we are trying to make sense of it."

In James's view there simply is not a pathway for conducting climate science in a legitimate and credible way, given the lack of verifiable data. While James has a basic sense of how scientists determine the earth's past climates, he does not find those methods to be credible:

> How do I solve this problem when I don't know what the average global temperatures were a thousand years ago? I don't know what they were two thousand years ago. I don't know what they were three thousand years ago. I know you can get some inkling by—you can get tree rings and layers of glacier and you can count the—it's not like we have absolutely no idea if it was a hard winter, this many winters ago. But that's not temperatures. And when you are talking about global warming you are talking about fractions of a degree and maybe a degree or two, and there's absolutely no way that we have that level of knowledge, so I conclude that you guys are blowing sunshine out of me. That's why I just don't buy it. Especially when the arguments are

that lame and they have obviously so much to gain if their programs are funded or if their control over me is given. They have a lot to gain and I just say, "Sorry, I'm just, I'm a denier."

The concerns and conclusions raised by James seem to arise from a lack of awareness of scientific techniques used to determine past climate data, including paleoclimatology, a field dedicated to determining climate records from hundreds to millions of years ago. To reconstruct past climates, scientists combine proxy records generated from a number of different ways, including tree ring data, rocks deposited by glaciers, lake and ocean sediments, ice sheets, corals, fossils, and historical records from ship logs and early and ancient weather observers.[33] The variety of data used in this process, and the interdisciplinarity and complexity of climate science overall, makes disagreements over the proper methods of data collection and analysis an inherent feature of climatology. Skeptics who are inclined to question the credibility of climate scientists use this complexity to sow and maintain doubt about the science, among themselves and in their conversations with others.

Some participants in our sample even go to the extent of claiming that climate data are being intentionally manipulated by scientists to support a consensus view. Mark, introduced earlier, agrees: "I think so much of it is based on false data." For Ben, a conservative with some college education, the data simply cannot be conclusive because there is not enough global coverage. He argues: "It'd probably have to be like a third of the world, to be conclusive. Like either true or false. And it would have to be in every hemisphere, every quarter of the world, northern and southern, east, west. You'd probably have to do a third of the land where it's heavily cited."

These perspectives indicate a lack of accurate knowledge about how climate modeling works.[34] Notably, they also demonstrate how the political polarization of climate change intensifies as education and science literacy increases, as social scientists Drummond and Fischhoff suggest.[35] Among our participants, those who are conservative leaning and have college degrees are more likely to question the veracity of climate science based on concerns about data and methods. With higher levels of education and science literacy, people who are motivated to protect their skeptic identity use their knowledge to further rationalize and justify their skepticism, to themselves and to others. Nick provides a good example of the lengths to which skeptics will go to engage in such "identity-protective cognition."[36]

Nick purports to having downloaded and personally examined a global historical climate data set from the National Oceanic and Atmospheric Administration (NOAA). Through his own research Nick has concluded that there simply is not a legitimate and scientific way to say anything precise about global temperatures across time and space:

> They can't say anything about almost anyone at that period of time and that is because at this period of time the US had five thousand weather stations. Canada, though, was number two, and they only had about five to six hundred weather stations, and the next one down below there is Australia with about one hundred or so. And then below that you have Russia, with a couple dozen weather stations or so. Everyone else, even the UK, who was a superpower at that time, has single-digit weather stations. So there just wasn't many weather stations to make a comparison."

Nick argues that for most regions of the world, climate data goes back only as far as the 1950s, when the military established weather stations and put "a lot of weathermen on the front lines to forecast the weather so that they could move the troops forward." But, even within the short period from 1950s onward, there have been many "relapses" in data collection, "because of budget cuts in the US." Nick's claims are inaccurate: weather stations have existed since the 1800s and were exponentially expanded beginning in the 1970s (due to the increase in airline traffic) and then again in the 1980s (with the advances in computing technologies).[37]

To further justify his arguments related to data inadequacy, Nick points to a specific precipitation study in Idaho where "they only used twelve of the weather stations," then continuing: "You can make a claim off of a sample size of twelve? It's one of those, I can understand if Idaho had twelve, so when I saw that, I went right to the data set and I pulled it up and I was like, How many does Idaho have?" Nick concludes: "Yeah, so it is one of those, that the data is not there to make a claim." Nick's views reflect a concern that climate scientists are not accurately following the scientific method. He claims that there are "flukes in how we collect data" which makes the data "unreliable" and "remarkably noisy."

Others, such as Henry, a Republican with a college degree, argue that even if you agree that the available climate data are sufficient for analysis, climate scientists still conflate correlation and causation, which delegitimizes their results. He explains: "Without questioning the science behind the data and just looking at the logic and rationalizing of the arguments that he has presented [referring to scientific studies cited by Al Gore], it is good as far as it goes. But the main problem that I am seeing is that he is just arguing a correlation and there is nothing causal." Henry considers Al Gore a representative and spokesperson for the climate science community. To Henry, Gore's inability to convince skeptics speaks to the unscientific nature of climate science, "and he never in anything that I have seen has demonstrated a causal relationship. And if you have someone in a scientific community, if you have someone who is obviously well versed in the scientific method and data and [is in the] scientific circle and conversations, who is very obviously avoiding and neglecting to demonstrate valid causal relationships, it is not by accident. It is not because he doesn't know better. So right there is an automatic red flag."

To Henry, Gore's representation of the scientific community is dubious, given "his own lifestyle choices that are in radical contrast with what he is saying." Henry concludes: "If they want to have the most legitimate influence, they need to go back to causal relationships. They need to go through the scientific method as legitimately as possible. And that includes showing the data, here is the raw data, identify facts and assumptions, and then use the facts to support their assumptions and demonstrate it through causal relationships. I have yet to see anyone within the discussion do that."

These various viewpoints suggests that skeptics' concerns about climate data cannot be easily dissuaded given the inherent uncertainty involved in how complex ecological phenomena are modeled. In fact, skeptics take particular issue with climate modeling, arguing that modeling does not constitute a valid form of science to begin with. Take Nick again. Nick questions the weighting techniques applied in climate modeling, arguing that scientists tend to "weight to consensus." To Nick, statistical weighting is deceptive, scientists are not fully transparent about weights, and they even use this method to explicitly manipulate the truth. He contends, "So weighting is one of those things . . . it's a statistical black magic box and if you understand what it is and how it works it is great, but even if you don't understand what it means, you can manipulate it to say whatever you want. So, you can make it say the truth, you can make it an outright lie." Nick questions other aspects of modeling, such as "removing outliers," which in his eyes delegitimize climate models and make scientific claims about climate change problematic.

Douglas, the biology teacher who expresses concerns about funding, does not agree with "so many assumptions that go into analyzing an ice core data set," for example. He feels "pretty confident in the last seventy years of data, the CO_2 [carbon dioxide] data is really convincing for the last seventy years," but he takes issue with extrapolating this data, which is a central element of climate modeling.

Skeptics who question climate modeling often cite the tendency for predictions to be too broad or to require adjustments as evidence of methodological shortcomings. For example, Greg, a white man from Boise, refers to an early UN Intergovernmental Panel on Climate Change (IPCC) report, which predicted temperature changes "between ten and thirty degrees." The IPCC "revised it a couple years later and it was like 'well actually it's more like five and twenty,'" which he views as evidence of methodological shortcomings. While Greg does not think climate scientists are "straight-up lying," the frequency with which they change their models and how they "overstress and stretch the boundaries of how good their information is" leads him to conclude that it's "close to perpetuating misinformation."

To Bob "there's just tons and tons of stuff that's neglected or glossed over" in climate modeling. He cites earlier climate models that ignored methane as a contributing factor, "which is insanely more powerful of a greenhouse gas than carbon dioxide." Because of this omission, prior work constitutes "poor science," which to him indicates that "there's other things that exist that they [scientists] don't discuss." In Bob's perception these omissions invalidate climate science.

Developing a climate model is a complicated process that involves identifying and quantifying earth system processes and setting parameters to represent initial conditions and subsequent changes at a planetary level. Once a model is set up, scientists use powerful high-capacity computers to repeatedly solve mathematical equations embedded in the model, including testing the model by running it from the present into the past to compare it with known climate and weather conditions that are known to have occurred and continue to occur. When a model performs well its results for simulating the future are also accepted as valid.[38] Given the complexity of this process, most lay people, including the skeptics we interviewed, may not fully grasp how modeling works. As a result, trust in the scientists who create and use modeling becomes a crucial factor that determines whether or not one agrees with the scientific methods employed in modeling or the conclusions reached. Where distrust prevails, skeptics continue to question the science around climate change, including referring to specific data sources, to question its credibility.

Exclusion of Outside Voices

Lastly, as part of the process of co-constructing scientists as a threatening out-group, skeptics argue that climate scientists prevent them from getting access in the scientific discourse. They argue that climate scientists exclude skeptics from civil discussion, they write in a way that is inaccessible to the general public, and they practice credentialing to exclude skeptical voices from scientific circles.

From the perspective of skeptics, climate scientists who agree with a consensus view are not willing to engage in debate or even civil conversation. They exclude and demonize people who hold contrarian viewpoints, including contrarian scientists. James succinctly summarizes the dangers of this perceived exclusion: "The scientific establishment has become an orthodoxy and they persecute [those who disagree]."

Consider Savannah, a Christian woman from southern Idaho who grew up in a working-class family, with the "philosophy that everybody works," "everybody had a chore." Savannah did not relish attending school, but she enjoyed "learning." After only a couple of years in public school she switched to homeschooling, which she enjoyed "because then I was able to just learn at my own pace without the added opinions of others." She identifies as a "socially awkward" person who did not fit in with her peers.

On climate change, Savannah argues that no matter how much self-learning and research she does on her own, nor how much she knows, she will always be ignored. "I had one year of college . . . I wouldn't exactly be considered a very credible source . . . But, regardless of whether I've had the studying or the college degree, that does not hinder me from reading the same subjects . . . Not only does it not hinder that, it doesn't hinder my understanding, either." Savannah argues that her lack of educational credentials does not preclude her from reading, understanding, and forming an informed opinion about the topic: "I could understand the subject far better than the person who has the credibility, if I liked it well enough or if I had a different way of learning." And yet, despite leaving school because she did not enjoy the "added opinions of others," she finds it unacceptable that people like her are excluded from science.

Skeptics view credentialling as a way for the scientific community to reject science that challenges the status quo. For example, Nick who is self-employed and has an associate degree in information technology, posits that "the problem with this is, I'm not someone with a PhD next to their name. So, it's one of those—Okay, so the hoops for me to get this published—it's huge." Nick argues that the only way he can get his independent research published in a reputable journal is by collaborating with someone with a PhD. He feels this process is unfair. His research should stand on its own merit and be welcomed into scientific circles.

During our interview Nick explained in detail the difficulties he has faced trying to publish research as an independent scholar, including the cost, which was "over a thousand dollars." Nick argues that his lack of affiliation with a research organization puts him at a significant disadvantage. "My options are to lie [about institutional affiliation], which, that's not going to be good. And so I just gave up on *Nature*." Given these credentialing practices and his exclusion from scientific discussion, for Nick "the only thing I can do at this point is really argue about it online with people."

Nick asserts that because the exclusion of contrarian viewpoints is so strong within the scientific community, which is "already on the side of 'dissent is garbage,'" he has struggled to write an abstract for an upcoming conference. He shares that "I had been having a hard time with the abstract that I'd been writing anyways. It was one of those—Okay, do I write the abstract so someone reads this first? Which means not telling them that this is global warming dissent. Keeping anything that might be—it's like, all this is a study of evaporation and precipitation and how it relates to global warming and not saying what the result is, not saying anything like that. Just that vague." From Nick's perceptive, such barriers are "unfortunate," given that outsiders are more likely to engage in objective work. "I don't have any money from oil and I don't have any incentive from the government to do pro–climate change studies. I'm a nobody who happens to know how computers work really well. So, uh, it's—it's unfortunate." All in all, Nick and other skeptics view scientific gatekeeping activities as deliberate efforts to prevent the entry of legitimate yet opposing findings into the scholarly discourse.

The perceived inaccessibility of scientific writing and the unavailability of free, scientific articles is also viewed by skeptics as further evidence of their exclusion from science. For example, Greg states that scientists "often write in a manner that's difficult to read." Scientific writing includes "a lot of hedging, there's a lot of 'may,' there's a lot of 'possibly,' there's a lot of 'should be', 'could be,'" which Greg calls "weasel words." As an engineer by profession, he argues that "a client doesn't want to hear 'on one hand, but on the other hand.'" In this way, the language used by scientists to capture uncertainty is seen as ambiguous and unhelpful by skeptics who are searching for specific answers.

Largely agreeing with this perspective, Jodie, a Republican-leaning Catholic from northern Idaho and the only person of color in our interview sample (she identifies as Latina), posits that climate change research is "hard to read." Scientific claims get muddled in the translation and dissemination to the public. "People are trying to interpret it. They say, 'This is what the scientist said,' even though it's not what he says." Jodie feels overwhelmed and unable to examine the original research by herself "because it seems to be a lot." Even though she claims to have tried, the research is not accessible to her because "you have to pay for the abstract or something like this. So, it's mostly what I can take off of articles and I don't have that off the top of my head."

While skeptics raise numerous concerns about perceived exclusions that reduce the credibility and legitimacy of science, the various gatekeeping activities in science are designed specifically to maintain its legitimacy and credibility. Take, for example, the gold standard of scientific publications, peer review. This process, particularly double-blind peer review, is designed such that meritorious work can get published and added to the extant literature regardless of who the author is and what their institutional background is. This is not to suggest that the peer review process is entirely error free, unbiased, and has no limitations, but rather that the viewpoints raised by skeptics indicate a lack of awareness of the purpose of these gatekeeping processes and a distrust of what they produce.[39]

Through this process of questioning scientists' integrity, skeptics create their social identity in parallel to the social category of climate scientists. Climate scientists are seen as an unethical and untrustworthy out-group easily influenced by nefarious motives and external forces. In contrast, skeptics construct themselves as victims, shunned for simply seeking the truth about an issue that they care about.

CONCLUSIONS AND IMPLICATIONS

The climate change skeptics we interviewed and surveyed have developed their self-identity as skeptics in opposition to climate scientists. In co-constructing climate scientists as an out-group, skeptics view them as unduly influenced by pervasive incentives, unable or unwilling to employ valid scientific methods, and resistant to public engagement through conversation and debate. Skeptics also allege that scientists marginalize those who do not agree with the consensus view on anthropogenic climate change.

Overall it appears that in developing and holding the climate change skeptic identity, skeptics view the scientific community as a threat, mainly due to the misalignment between the narrative of human-caused climate change and skeptics' beliefs on the issue. To skeptics, climate scientists are the category of people most responsible for this identity-threatening narrative, solidifying the us-versus-them dichotomy between themselves and scientists. The sense of threat skeptics experience from the scientific community leads them to derogate scientists, which in this case leads to a particular distrust of climates scientists and a dismissal of climatology in general.

In fact, climate scientists are one of several groups viewed as threatening out-groups by skeptics. Other social categories, such as environmentalists, liberals, and any mainstream media that promotes or acts on the consensus scientific view on anthropogenic climate change, are part and parcel of this us-versus-them dichotomy in a skeptic's identity construction. To skeptics, these others exclude and marginalize their views while perpetuating the insufficiently scientific or outright false narrative of climate change.

Recognizing that the contention between climate change skeptics and climate scientists is rooted in an oppositional identity dynamic has important implications for communicating climate science. Beyond common cognitive traps, such as confirmation bias, the us-versus-them identity dynamic will cause any messaging from climate scientists to be met with both skepticism and a deeply rooted, emotionally laden form of distrust. For messages to be effective they need to be presented by more trusted or identity-aligned sources (e.g., conservative news, fellow seekers, libertarian politicians). Other methods that might help reduce distrust and enhance constructive intergroup dialogue, such as making peer-reviewed climate science openly and widely available and written to be easily understood, should also be considered when communicating the scientific consensus with skeptics.

We now turn to explore the nuances of the skeptic perspective and elaborate on the implications of the skeptic identity on climate communication. Future chapters examine the cultural context within which US climate change skeptics operate, how their opinion-based, stigmatized social identity is continuously shaped and reshaped, and which actors are involved in construction of the skeptic identity, covering identity categories and topics ranging from political elites, religious leaders, and conservative media, including those who run outright climate disinformation campaigns.[40]

three

RELIGIOUS IDEATION AND CONSPIRACY ADHERENCE

JODIE IS A COLLEGE STUDENT in her early twenties. She was raised as a Republican and spends a great deal of time with fellow conservative Catholic Latine families. Jodie was homeschooled for part of her education, using a religious curriculum. Her homeschooling consisted of what she called "a creationist textbook," which discussed climate change. This text has largely informed her current views on the topic. As a result, she believes that the earth's climate is changing but that it is not "manmade . . . it's natural." In her mind humans simply cannot have a "global, long-term" impact on the climate because Earth is God's perfect creation.

Jodie's faith predominantly shapes her views on climate change. She believes that God is in control and humans are arrogant to think they can have an impact on the climate. "It's not something we can affect in a global, long-term level . . . It's a natural thing and not something that we could have prevented." Despite this belief, Jodie does think that certain environmental problems, such as pollution, are worthy of concern and that humans can have "a short-term, kind of like a smaller-scale" impact on what she sees as more regional environmental issues. She is clear in her beliefs that "deforestation is bad" and that "cleaner air . . . will help in the future."

Jodie's religious beliefs also impact her understanding of science. For Jodie, science is unable to reveal the secrets of God's plan. She says: "In terms of trying to be Mother Nature, I don't think that's possible. I don't think we have the resources to do that. I don't think we have the information to predict that what we can do is actually going to make a difference." When reflecting on climate science and scientists, this idea of projection is very difficult for her to comprehend. She questions, "What doesn't make sense to me is how they can predict what is going to happen to the earth so far in the future. Like, one hundred years from now."

Jodie's worldview also shapes her engagement with news media. When asked what information sources she most trusted, Jodie responded, "I don't really." Yet she finds some comfort in online spaces in which people can find information "from both sides." She herself relies on "almost every news source" but especially likes "political podcasts" and "videos on YouTube."

Jodie's confidence that Earth is a gift from God mitigates a strong emotional response to climate change. She reports feeling "curiosity" when encountering stories about climate change, not anger or sadness. "I don't feel angry or sad. I'm not angry like, 'We are destroying our world, we need to do something right now' . . . it's more, I think, a neutral thing."

Through our project it became clear to us that Jodie's religious ideology shapes the manifestation of her identity as a climate change skeptic by affecting her perception of climate change, climate science, and the news media. It also influences her emotional responses to the topic of climate change. In truth, one's core ideologies influence, and even create, the realities through which they experience the world. We now look at two ideologies that both participants and prior scholarship identify as producing variations among climate change skeptics—conservative religious beliefs (i.e., religious ideation) and the perception that climate change is a conspiratorial hoax (i.e., conspiracy ideation).[1] We first explore existing scholarship on how ideologies operate and then examine these specific belief systems as they influence the skeptics in our sample.

IDEOLOGY

An ideology is a set of ideas to which one adheres that contextualizes their experiences of the world and in turn shapes their thoughts and behaviors. For adherents, ideologies are not "true"; rather, they construct experienced reality.[2] Ideologies are illusory, powerful, but invisible ways of seeing that build on "a false consciousness" that both shapes and distorts understanding of oneself and the broader world.[3] An ideology is not simply an opinion—though opinion is certainly shaped by ideology—but rather a system of ideas with its own internal logic that is used to make sense of all social phenomena and experiences.[4]

Ideologies are learned through experiences with others. Starting early in life individuals are primed to see the world in a particular way, in line with the culture into which they are born. Parents, teachers, and friends serve as teachers of ideology and help create the very reality to which one subsequently adheres.[5] Political, religious, and cultural beliefs create and reinforce the reality that one experiences. This may suggest that ideology is a very personal set of beliefs, yet there are some ideologies that are so pervasive that they create the basis for the institutions of society and the very ways that individuals behave within them.[6]

Ideologies also shape, intersect with, and are influenced by the identities to which an individual ascribes. Consider religious identities. Sociologists Lisa Pearce and Arland Thornton have looked at the relationship between maternal religious identity and the ideologies held by families. They find that religious identities and associated belief structures serve to shape the foundational ideologies of the family.[7] This is not surprising, as the identities and the ideologies to which one adheres are so interconnected as to be difficult to evaluate or assess apart from one another.

In the case of climate change skepticism, any number of ideologies may weave together to create the specific set of beliefs held by an individual. However, certain shared ideologies exist that profoundly, and with great complexity, shape the worldviews, logic structures, and behaviors of their adherents. Here we explore two such sets of belief structures—specific religious attitudes as they impact climate change and the perception that climate change is a hoax—to better understand how ideology shapes climate change skepticism.

Religious Ideation

In the study of social phenomena, religion serves as an important variable for social scientists. For climate change skepticism specifically, scholars often consider the impact of "religiosity," measured as the frequency of religious service attendance (by some scholars, such as McCright and Dunlap) or self-reports (by others, including Jaesun Wang and Seoyong Kim).[8] In our survey work we found that specific religious beliefs serve as powerful predictors of a set of climate change skeptics' beliefs: the perception that climate change is a hoax, distrust in science, and support of environmental policy. We relied on four survey measures to reflect specific religious ideation: that climate change indicates God's will (to which 18 percent of our sample agreed); that climate change is the end-of-days as predicted by the Book of Revelation (13 percent); that climate change is punishment for human sins (9 percent); and that humans are meant to rule over nature (39.4 percent).[9] The distinction between religiosity and religious ideation is important to note, since religious ideation reflects the ways that the specific ideologies within a religious community shape personal reality and understanding of climate change.

Religious ideology, like other ideologies, constructs a reality on which adherents base their lives.[10] Religious beliefs serve as a foundation for people's interpretation of new information and their behaviors, attitudes, and even interactions with others. When considering climate change skepticism specifically, religious ideology impacts skeptics in powerful yet complicated ways. For some, such as Jodie, religious beliefs lead them to feel a strong sense of responsibility for certain environmental issues, both because Earth is a "gift from God" and because doing so is part of the Golden Rule. For others, Earth as a gift from God justifies human exploitation of natural resources.

Douglas, introduced earlier, is in the first group. Coming from a family of five relatively moderate, democratic-leaning individuals, Douglas has always stood out ideologically from his family. In fact, for much of his life he struggled to find an ideological home. He tested out the Republican Party, the Green Party, and several other ideologies before ultimately settling on "apoliticalness." While not politically affiliated, Douglas is deeply religious and is currently employed as a teacher at a Christian school. He "see[s] [his] job as a Christian is to love God and love my neighbor." For Douglas, this extends to environmental stewardship. He explains: "When we share resources like air and water, you know, how can I love my neighbor . . . what I do over there in the air is going to impact the air here . . . what I do in Idaho with the water is going to impact people in Washington State." As a result, Douglas feels that we need both regulations and free market controls to enforce environmental stewardship. He argues, "I definitely think there needs to be some type of regulation for air and water pollution, and when it's cheaper to pollute, that infuriates me."

Douglas is confident that he is not alone in these beliefs, and our interview and survey data show this to be true. Fifty-six percent of the skeptics we surveyed believe that "the Golden Rule requires that we try to not pollute the earth." To that end, we find that skeptics show strong levels of concern about environmental issues such as water and air pollution, habitat destruction, and support for renewable energy (discussed in detail in the next chapter).[11] Douglas contextualizes these perspectives as "the greening of the Evangelical Church."

While evangelical and fundamentalist Christians are more likely to deny climate change and, according to some studies, less likely to hold or enact pro-environmentalist beliefs and behaviors, there is an active environmental movement within Evangelical Christianity. Perhaps most recognized is the work of Katherine Hayhoe, an atmospheric scientist who is also an Evangelical Christian and advisor to the thirty-thousand-member Young Evangelicals for Climate Action.[12] Social scientist Laurel Kearns argues that the challenge the environmental movement faces within conservative Christian organizations is their global message.[13] They argue that evangelical faiths have largely become intertwined with American capitalist values and the idea of God as all-powerful and all-knowing; these foundational beliefs make the solutions proposed by environmentalists difficult to find traction. This combination of religious identity (evangelicalism), associated capitalist values, beliefs about environmental stewardship, and climate change skepticism, create tension for skeptics such as Jodie and Douglas. The resulting cognitive dissonance often leads to one either changing identity or bending associated beliefs to accommodate conflicting identities (see chapter 8). However, a third option, one less commonly seen, is to simply live in the tension.[14] Douglas seems to have embraced this direction, despite the sense of further marginality it produces for him in that he does not feel like he fits in fully with skeptics or with environmentalists.

Yet Douglas and his fellow religion-inspired stewards are not the only skeptics influenced by specific Christian beliefs. Another group of religious skeptics cite their faith as evidence of their general apathy. While fewer in number, these interviewees revisited an argument commonly expressed among religious skeptics as to why climate change does not exist: that God would not let it. James serves as a good example of this framework. James is the pastor of a fundamental Calvinist church. He has a reputation in his community of self-aggrandizement and during his interview he recalled school as failing to challenge him. At the root of his faith, James explains, is the belief that "the planet was exquisitely designed as a gift for us." He argues that God "made [it] for us and we were made for it." As a result, he contends, "God is not trying to kill us all off. If He were trying to do that, we'd be dead." He goes on to apply this logic to environmentalism, or his lack thereof. He states: "God gave us this world as a place to live. As a creationist I believe that God is going to take better care of our habitat as an ongoing gift."

In our examination of the relationship between skeptics who hold specific religious beliefs and associated levels of environmentalism, we find that skeptics who believe that humans have a right to dominate over nature are less likely to be concerned about environmental issues. Consider, for instance, James's argument. James believes that the planet is a gift from God to humans; that humans can and should use Earth for their benefit. Though he does not fault his peers for having a sense of environmental stewardship, he lacks a sense of concern about environmental issues such as sea level rise, heat waves, and melting ice caps. James does not believe they are even occurring. Yet he does have some concerns about pollution. James is aware that air and water pollution can harm human populations, so that, while he does not want to "have air pollution like it was in the sixties," he believes that environmental action on these issues is "the overreach of the Obama years" and he argues against state involvement in managing pollution.

James reflects what scholars have identified as the lower levels of environmentalism found among Evangelical Christians, a perspective known as the "Lynn White Thesis."[15] Religious scholars contend that people who interpret the Bible literally are less likely to support environmentalism when it is framed in competition with economic interests. They are less likely to engage in pro-environmental behaviors, be willing to pay higher taxes for environmental protection, divest from high-polluting companies, and vote based on environmental issues.[16] In short, they are less likely to be concerned about the environment.[17]

Whether they are pro-environmental or apathetic in nature, religious organizations, ideologies, and identities serve as significant motivators among adherents. They influence a believer's perception of right and wrong and accompanying "moral duties."[18] But, they also operate in a broader ideological context, which likely shapes their attitudes toward environmentalism. Sociologists Darren Sherkat and Christopher Ellison, for example, find that the political ideologies that underly different religious sects have a profound influence on an adherent's environmental behaviors and perspectives. They demonstrate that religious perspectives can shape pro-environmental or anti-environmental perspectives and activities, even though the divide largely reflects the political ideologies that correlate with various religious sects. Among climate change skeptics in our surveys, for a slight majority of religious skeptics (56 percent), religion inspires a sense of environmental stewardship, but for a significant minority (39 percent), it moves them toward apathy.

Our interviews suggest that whether a skeptic's religious views lead to stewardship or apathy might be an outcome of a belief that underlies the connection they make between climate change and religion. Skeptics who believe that climate change is a sign of the apocalypse or the Biblical end-of-days, skeptics who believe that God is in control, and skeptics who interpret the Bible literally all hold fewer pro-environmental views than skeptics who do not adhere to these beliefs.[19] Consider Karen's thoughts on this. Karen is a white woman living in northern Idaho. Growing up, her family took their Latter-day Saint faith very seriously. Karen was homeschooled and primarily interacted with other conservative Latter-day Saint families. Karen does not believe that the earth's climate is changing. In fact, she contends that it is useless, even arrogant, for humans to worry about climate change:

> God has a plan for this world and if He chose to wipe it out by global warming, then He could! If he wanted to . . . He's not going to be like, "Oh, shoot! These humans that I put on the planet that I made are fucking it up! What the hell?" No. You know? I feel like we're giving humans too much credit for the larger picture . . . If you believe in God, I think you have to give Him more credit . . . We are part of a bigger plan that may or may not affect a larger picture . . . I think we're just giving ourselves too much credit that we think we could change the world by driving a car.

Karen's underlying religious beliefs—specifically the belief that God is in control—directly informs her apathy toward the environment and toward climate change. Because God is orchestrating events on Earth, in her vision she need not worry about human action or carbon emissions.

Among climate change skeptics, specific religious beliefs, or religious ideation, significantly impact their perspectives on climate change as well as environmentalism more broadly. Participants such as James and Karen believe that Earth was made by God for humans. This perspective shapes their sense of responsibility about environmentalism. In contrast, some religious skeptics like Douglas feel a sense of stewardship rooted in a foundational religious belief that one must care for their neighbor. Regardless of which camp skeptics fall into, the foundational religious ideologies to which they adhere shape their broader perspectives on one's role regarding climate change and environmental preservation.

Conspiracy Ideation

Like religious ideologies, underlying political ideologies also shape the realities of climate change skeptics. Conspiracies around climate change typically take the form of a story in which a person or a group of people with relative social power have orchestrated a conspiracy to gain power or control in society. Adherents of this belief structure are more likely to be men, to attend church, and to identify as politically conservative.[20] Conspiracy theorizing around climate change is so prevalent among skeptics that in their *Conspiracy Theory Handbook* psychologists Stephan Lewandowsky and John Cook posit that deniers' most common response to information about climate change is conspiratorial in nature.[21] We do not find a level of adherence to conspiracies to be as high as Lewandowsky and Cook (40 percent): only 25 percent of the skeptics in our survey agree that climate change is a hoax.[22] Regardless, the conspiracy perspective is disproportionately reflected in news reports on skepticism and serves as an important driver of thought on this issue.

As with other ideologies, conspiracy ideation serves several purposes for adherents. Certainly, it functions to create a shared reality through which new information is processed. It also serves as a simplification of social problems, which may mitigate certain strong emotions. In fact, believing in conspiracies can have important emotional consequences. People who accept conspiracies tend to have lower levels of self-efficacy than nonconspiratorial people and experience heightened levels of anxiety, fear, and paranoid thinking.[23] They appear to accept conspiratorial stories about situations that are highly complex, around which there is some level of ambiguity, and which evoke strong fear-based emotions. Conspiracy theories tend to imbue certain powers in a single entity (in the case of climate change, examples include "global elites" or "Al Gore"), encapsulating the cause of negative phenomenon into a specific identifiable agent. Like religious ideation, conspiracies can serve to "fill a need for certainty, control, or understanding, filling gaps in knowledge and offering a coherent elucidation of difficult events."[24] In so doing, conspiracy adherents may gain a sense of control over a frightening phenomenon such as climate change, thereby reducing their fear, anxiety, and worry.

To illustrate this point, consider Blake, a Libertarian from southern Idaho. Blake believes that a reduction of volcanic activity and concurrent increase in solar activity is the cause of what he sees as a short-term increase in global temperatures. Instead of accepting climate science, Blake contends that the United Nations, seeking tyrannical control over global citizens, has manufactured climate change to enact certain authoritarian policies. As evidence for this perspective, Blake opens his interview by arguing that all climate reports are written by the United Nations, an organization dominated by elites: "They are from different nations and they are from the elites of those nations, so you have people who have become dominant who get into these roles . . . everyone who is approving a final draft of a publication of that magnitude is going to be an elite." One such elite Blake specifically names is Al Gore, "who is, by every definition an elite, child of a billionaire, former vice president at the time, member of the Council on Foreign Relations, involved in these international discussions about what are we going to do for going forward." Blake believes that the United Nations has manufactured the discourse around climate change to both control oil reserves around the world and to violently restrict human reproduction. Blake takes a surprising, traditionally liberal approach in his critique, suggesting that UN programing undermines local cultures and is therefore unethical. He states:

> The decision was made by a small group of Western elites that we're going to get all of the globe's population down, or growth down. Like, how are you going to do that? You're going to have to tinker with people's cultures all over the world. You can't just do that, right? So, then you have serious pushes towards women's education, which sounds like a really good thing, except that the education has to fit certain criteria for it to be good in the eyes of these Westerners. So they are essentially driving out cultural aspects that are not sustainable in their definition of sustainable.

Blake continues by arguing that the United Nations and its puppet media institutions disguise more violent programs to restrict populations:

> In the Middle East, the boys are getting killed and then some girls get captured and then released and there's no media on it. No media on the boys getting killed or the girls getting captured and released. Then they kill the boys, and these are like, you go in, you take the boys, you kill them, you take the girls, you send them home, and you say 'Go and be a wife and mom'. . . That's really suspicious. If we're really concerned about terrorism we'd be concerned about selectively murdering boys, not selectively preventing girls from getting a Western education. Pretty suspicious. So, anyway, there's a push for global education for sustainability, which I think includes reducing population growth.

While to those who believe in climate science this perspective appears out of line with reality, Blake is not alone in his belief that the UN and others are murderous villains in search of control. Bill, a Libertarian from northern Idaho, makes a similar claim when he contends that "it's basically more about the idea of controlling the population, really . . . There was an organization called Zero Population Growth . . . They basically think that the earth has limited resources and the fewer people we have around the better . . . It gets bent down to power and control. Basically, trying to control the people to have them do whatever the politicians want, the leaders want." Bill has spent all his life in the inland Northwest. He graduated from college and has held jobs in a variety of fields since graduating. After marrying his wife, Bill converted to the Church of Jesus Christ of Latter-day Saints. Based on his personal experiences, upbringing, and religious beliefs, Bill believes that the climate is changing but as a natural result of "sunspot activity and . . . an increase in solar radiation and the shrinking of the polar ice caps on Mars." Like Blake, Bill blames the United Nations as one of the primary drivers of what he sees as the manufactured climate change discourse.

Not all skeptics who believe climate change is a hoax think that the United Nations is the primary entity aiming for world dominance. In fact, in our survey of one thousand skeptics we found that about 67 percent believe "global elites" are involved in advancing the climate change hoax; 53 percent state that the UN is responsible; 22 percent cite the Council on Foreign Relations; and 65 percent hold Al Gore responsible.[25] Regardless of which entity they perceive as advancing the myth of climate change, the theme of control and manipulation is broadly shared among this subset of skeptics. Of skeptics who believe that climate change is a hoax (roughly a quarter of our survey sample), 73 percent believe the hoax is orchestrated to gain power and 66 percent believe the climate change discourse is put forth to make money.[26]

Lee elucidates the perspective that power is the driving force of the false climate change narrative. Lee is younger than most others we interviewed. A senior in college, he questions the veracity of climate change due to the discursive shift from first referring to it as "global warming" and then "climate change." He believes this shift is due to an intentional manipulation on the part of global elites. "They're just changing it to try and push stuff into it . . . I just saw it as more of an agenda." A Republican from northern Idaho, Lee likened the climate conspiracy to the show *House of Cards*. He explains:

> Have you watched *House of Cards* at all? It's a great illustration of how some people are motivated by money and just getting money, but other people are motivated by power and being able to have a say in what, in how events turn out, and how people live their lives. And so I think that one thing that people have to gain by spouting that sort of information is power over how people think and how they vote and how they interact with their environments . . . sort of indoctrinating our generation with this . . . Climate change is just a tool that people are using to get reelected, so that they can stay in power and keep making money . . . I just think there's a lot of money to be gained when you can, um, control something, a viewpoint, and have a monopoly on it.

Lee explains the hoax by arguing that the politicians who drive it are after power and money. Money is the second-most cited explanation for the perceived conspiracy in our sample; 66 percent agree that money is a potential driver. Many conspiracy-inspired skeptics believe that governments or the United Nations advance the hoax of climate change to increase their personal and institutional wealth.

Consider as an example Mark, the Libertarian from northern Idaho. Like other conspiracy-minded skeptics, he believes that climate change is a hoax contrived by the United Nations to attain global domination. "It's about control and power and money and, you know, milkin' the people." Mark believes that the UN and its cronies within the American government have falsely constructed climate change to "get the masses scared so they'll accept carbon taxes . . . to get people more submissive, I think . . . The powers that be want to dictate what the narrative is and these guys gotta go along to keep their funding and stuff and whether this business of the carbon tax is just another tax that the elite'll make a lot of, well, Al Gore, but the elite'll make a lot of money on that."

We find that adherence to the belief that climate change is a hoax also correlates with lower concern about environmental issues. Skeptics who believe climate change is manufactured are approximately 35 percent less likely than their skeptic peers to be concerned about sea levels rising, 28 percent less likely to be worried about deforestation, and 27 percent less likely to be concerned about an increase in heat waves. Even for topics less commonly associated with climate change, such as plastic in the ocean, skeptics who believe climate change is a hoax are approximately 24 percent less likely to express concern than those who do not agree with the existence of a climate hoax.[27]

Blake (introduced earlier) does hold some concerns about certain environmental issues that affect human health, namely, water pollution and air pollution. He also worries about "dust and smoke" but contends that government policy is unlikely to touch these issues: "Neither of those are going to get regulations to clean them up." He does not hold concerns about environmental issues that are often associated with climate change in news reporting. For instance, he believes that sea levels are rising and that it is due to warming, but "that warming is not correlated with CO_2, so you will see this [sea levels rising] until the global temperatures go back down." Blake holds similar feelings about extreme weather patterns arguing that "that data just doesn't exist. We just don't have it."

These belief structures—religious and conspiracy ideation—operate with and through one another. People who hold religious ideologies are likely to follow conspiracies that align with those religious beliefs. For example, research by Anna-Kaisa Newheiser, Miguel Farias, and Nicole Tausch finds that people who adhere to New Age ideologies are more likely than those who are Christian to believe in the Da Vinci Code conspiracy, while Christians are more likely to believe in medical and political conspiracies.[28] In analyzing our survey data we find that skeptics who believe that climate change is God's will, that it indicates the end-of-days, or that it is punishment for sin, are significantly more likely than skeptics who don't ascribe to these beliefs to also believe that climate change is a grand conspiracy and a fictitious "hoax."[29]

As with religious-ideation, conspiracy-ideation appears to set apart the skeptics who believe that climate change is a hoax from their peers in meaningful ways. This foundational belief impacts their understanding of and concern for certain environmental issues and results in what we call a "conspiracy gap in concern for environmental issues."[30]

ADDITIONAL EFFECTS OF IDEOLOGIES

So far we have explored how religious and conspiracy ideologies impact skeptics' understanding of climate change and associated environmental beliefs. While these trends are significant, they are not the only ways that adherence to certain ideologies are associated with other beliefs. In fact, ideology serves as the foundation for creating a worldview and shapes values and behaviors in ways that remain in line with knowledge systems.[31] Ideology also determines emotional responses to social stimuli.[32] When considering the impact that religious ideology and conspiracy adherence exerts on a skeptic's trust in science and media as well as their emotional response to climate change, we find a measurable and meaningful "gap" based on adherence to these two foundational ideologies.

Ideation and Trust

In chapter 2 we examined the relationship between climate change skepticism and trust in science. We find that while religiosity does not directly correlate with trust in science among skeptics, specific religious beliefs are associated with a greater distrust. Specifically, skeptics who believe that climate change is God's will, is part of the end-of-days as predicted in the Book of Revelations, or is punishment for sins have significantly lower levels of trust in science than do their skeptic peers holding different religious beliefs; skeptics who believe that humans have a right to rule over nature also have significantly lower levels of scientific trust.[33] People who hold these religious beliefs are also less likely than their skeptic peers to trust science-based organizations such as the Environmental Protection Agency (EPA), the National Oceanic and Atmospheric Administration (NOAA), the National Aeronautics and Space Administration (NASA), and others who report on climate science such as weather forecasters and climatologists.

For example, James thinks that scientists have "been educated into a very untrustworthy scientific system." He argues that "they are deceived" and that, as a result, the work they produce is untrustworthy: "He might write a book on climate change and I don't believe a word of it." In particular, James is distrustful of computer modeling, which he calls the "GIGO principle, which is 'garbage in, garbage out.'" He does not trust the data that scientists enter into their models or the results that the models produce.

Regarding conspiracy ideation, skeptics who believe climate change is a grand conspiracy are significantly less likely than other skeptics to trust climate scientists, other scientists, doctors, and even weather forecasters.[34] Lee, for example, believes that climate scientists are negatively influenced by money to such a degree that their work is unreliable. "I think that people are greedy and that everybody has a price, but, you know, whether that manifests itself in skewed data, as a norm, I would say it's the majority of the time." Blake share's Lee's concern with coercive funding practices, especially when it intersects with his broader distrust of the government. Blake believes that any scientist "receiving federal money is suspect."

Ideology also shapes trust in news sources and the media, but in distinct ways. In analyzing our survey data we find that while religious ideation does not significantly affect trust in media, conspiracy ideation is associated with lower levels of trust in mainstream media (e.g., NPR, CNN). However, skeptics who perceive climate change to be a hoax are still more likely to trust the Fox Network more than their skeptic counterparts.[35]

Blake, for instance, argues that mainstream media is untrustworthy "on three different levels." He contends that there is a "governing body tell[ing] the news media what they can report on some politically charged issue" and that "the mainstream media decide what they are going to put out—decide what their stance is." Blake believes that all mainstream media sources, "except for *Fox News*" are "certainly pushing climate change." As a result, he and others who believe that climate change is a hoax, reject mainstream news: "I do not watch or listen to mainstream media. The system is broken." Instead, Blake relies exclusively on internet searches for his information.

Lee shares a similar perspective, saying, "The media and I aren't on good terms right now." Like Blake, he contends that mainstream media sources alter content for ulterior motives, though where Blake cites political directives, Lee sees the goal of media sources as monetary profits. "They're in the business of viewership, so they're going to print and write and make videos on stories that are gonna get views." Lee relies on social media instead. Even though he recognizes that social media is shaped by the person sharing information, Lee thinks that "overall social media is pretty balanced because there's a balance between conservative and liberal ideals."

Ideation and Emotions

Beyond the concept of trust, ideologies further impact the emotional experiences of adherents. Indeed, we find that adherence to religious ideology increases certain emotional responses to climate change, while conspiracy ideology serves to mitigate some emotions that skeptics have toward environmental issues and climate change.

For example, skeptics who believe climate change is God's will, the end-of-days, or punishment for sins are more likely than their peers to express dread or sadness regarding the subject of climate change.[36] This may indicate a higher level of belief that climate change impacts human society compared to their peers. For example, consider Savannah. Savannah was raised as a member of the Church of Jesus Christ of Latter-day Saints. While she has since left the church, she still believes that when humans rely on science, they are "playing God" and "trusting that you're smarter than God." Savannah does believe that humans have a degree of responsibility for taking care of the earth, but she also believes that climate change is naturally occurring. Nonetheless, climate change brings Savannah great sadness, not so much because of a conceivable impact on human livelihood but because of the potential extinction of animal species. She states: "It just would be a really sad world for my children not to be able to experience that." Savannah believes that the earth will survive, but that "there's danger in losing certain species of animals, which would be a tragedy, in my mind."

In contrast, skeptics who believe that humans have a right to rule over Earth are less likely than their peers to have negative emotions about climate change. This is in line with our findings regarding the belief that climate change is a hoax. Skeptics who believe climate change is a hoax are significantly less likely than other skeptics to experience worry, dread, sadness, or grief when thinking about climate change and significantly more likely to feel calm. There is no significant difference between these groups in their feelings of disgust.[37] James, for example, says that his emotional response to discussion of climate change is "generally the equivalent of rolling my eyes."

These findings are not particularly surprising. It makes sense that, if one not only rejects that climate change is occurring but also constructs a counternarrative (that it is a hoax) to explain the behavior of climate scientists, politicians, and other leaders, they would experience a reduction in negative emotional responses to climate change. As noted earlier, conspiracy ideation appears to act as a sort of psychological tool that reduces feelings of anxiety, fear, and worry.[38]

Blake, who believes that climate change is a hoax orchestrated by "elitists," does not have strong emotions about the phenomenon. He states: "When I think about humans causing global warming through CO_2, I feel no emotion." However, he does feel upset about what he views as a conspiracy and "the way that the agenda is being propaganda, it causes emotional distress in the same way that other things that are being propaganda causes emotional distress." As with trust in science and environmental attitudes, adherence to conspiracy and religious ideation shape skeptics' emotional responses to climate change.

CONCLUSIONS AND IMPLICATIONS

While climate change skeptics may share a common sense of identity and a specific view of climate scientists as an out-group, certain concurrently held ideologies shape an individual adherent's degree of skepticism in such a way that a meaningful distinction can be drawn between them and their peers. In other words, climate skepticism is nuanced and complex. Adherence to certain religious and conspiratorial beliefs shape skeptics' understanding of climate change, their sense of environmental stewardship, their trust in science and the media, and their emotional responses to climate change. This ideological "gap" is substantial and can be measured through statistical analysis, suggesting that ideologies can manifest as different skeptic "types" (e.g., conspiracy adhering skeptics and nonconspiracist skeptics).[39] This finding has particular implications for the success of different climate communication strategies. For instance, researchers have found that Christians are more receptive to anthropogenic climate change when the messaging includes appeals to Christian values of environmental stewardship.[40] Relatedly, media scholar Matthew Nisbet argues that to "break through the communication barriers of human nature, partisan identity, and media fragmentation" we need to frame messages in ways that resonate with specific audiences.[41]

Because climate science messaging is rarely targeted to specific skeptic "types," despite skeptics' ideological gaps, it is still meaningful to examine how a skeptic identity operates across concurrently held ideologies within the broader category of skepticism. Thus we turn to an investigation of how climate change skepticism is associated with concern about environmental issues and support for pro-environmental policies.

four

PRO-ENVIRONMENTALISM AMONG SKEPTICS

BEFORE DONALD TRUMP WAS ELECTED to the presidency, he made his beliefs about climate change clear. In 2012 he famously tweeted, "The concept of global warming was created by and for the Chinese in order to make U.S. manufacturing non-competitive." A year later he erroneously conflated the weather with the climate, advancing his claim that climate change is a conspiracy by tweeting, "Ice storm rolls from Texas to Tennessee—I'm in Los Angeles and it's freezing. Global warming is a total, and very expensive, hoax!"

Though the strength of this conviction appears to have ebbed once he took office (when he subsequently claimed to no longer believe climate change is a hoax but remained unconvinced that it is caused by human activity), Trump's policies were largely harmful to the minimal progress that had been made in the United States in addressing climate change. Upon taking office, Trump promised to retract the climate initiatives implemented by the Obama administration and infamously withdrew the United States from the Paris Agreement. The Trump administration eliminated funding for climate research, cut the budget of the EPA (Environmental Protection Agency), and completed rollbacks of over seventy environmental protections, including those regarding air and water pollution, toxic waste, vehicle fuel efficiency, and many more.[1]

Despite his anti-climate rhetoric and actions, Trump claims to be an environmental steward, arguing that his administration "made it a top priority to ensure that America has among the very cleanest air and cleanest water on the planet. We want the cleanest air, we want crystal clean water, and that's what we're doing."[2] Former Vice President Mike Pence, in his 2020 debate with Senator Kamala Harris, falsely claimed that under the Trump administration, "our air and water are cleaner than any time recorded," when in fact the Trump administration's regulatory rollbacks have reduced both air and water protections.[3]

Trump is well-known for hyperbolic and erroneous claims (according to the *Washington Post* fact checkers, he lied over sixteen thousand times during his first three years in office), but his rhetoric on climate change and the environment may be intentional.[4] Trump has been manufacturing the message that he does not believe in climate change. Though he waffles both on this position and the strength of his conviction, he consistently rejects that humans are contributing to climate change. His shifting message mirrors a concurrent shift in overall US climate change skepticism over the same period.[5] At the same time, he proclaims to care about air and water pollution. While this seems contradictory, given the direct relationship between air pollution and climate change, our research finds that this apparent disconnect is shared by many climate change skeptics. In fact, Trump's lip service to caring about pollution reflects a deep concern many skeptics have regarding air and water quality.

While climate change skepticism is an identity, group members do show variability in some belief structures. In this chapter we examine this variability and share their primary environmental concerns (pollution and trash, habitat destruction and animal species loss, and renewable energy), showing how skeptics disconnect these specific concerns from climate change. We further examine the policies they support to ameliorate the problems they cite and the implications of these findings for climate change communication and policy.

POLLUTION AND TRASH

The primary environmental issue of concern to skeptics is pollution. Nearly every person we interviewed agrees that pollution is worthy of concern and explains their personal fears regarding air pollution, water pollution, soil health, and waste. It is important to understand the context in which these concerns arise and why skeptics do not see pollution as connected to climate change. We also present examples of what social scientists call "negative cases"—the two participants in our interview sample who were not concerned about these issues.

Among the people that took our survey, 62 percent said that they are concerned about air pollution.[6] Indeed, since the Trump administration took office in 2016, US air quality has steadily declined, according to the EPA.[7] Air pollution, which results primarily from the production of energy (coal, gas), is monitored and regulated under the Clean Air Act, originally passed in 1970.[8] Air pollution can have numerous negative health impacts, including damage to the eyes, throat, and lungs. Problems are especially pronounced for people with asthma or other breathing disorders. Some air pollutants are also associated with cancer, liver disease, immunological weakness, nerve damage, brain damage, harm to the endocrine system, and reduced reproductive function. In addition to directly harming human health, air pollution also has a negative impact on the climate. Many pollutants are considered greenhouse gasses, which "by trapping the earth's heat in the atmosphere" increase the overall temperature of the earth's climate, which is further associated with negative impacts for humans and animal species.[9]

Most people in our interview sample, including David, the biology teacher at a religious college, share this perspective. David has lived in a variety of places, including on the East Coast and several western states. David believes that climate change is occurring but does not think that humans cause it. When we asked David about pollution, he agreed that pollution is a concern. He wants to "keep the air clean, keep the water clean." He ends his argument saying, "It's not like the conservatives want to breathe dirty air!"

Just as no one wants to breathe dirty air, people—including skeptics—do not want to drink contaminated water. Our survey found that 62 percent of skeptics are concerned about declining water quality.[10] One of the people we interviewed, Pam, expresses concerns representative of other skeptics. Pam was born and raised in a small town in southern Idaho. She believes that climate change is a natural process and that humans do not contribute to the phenomenon. Yet, she argues, "we can do some things to kind of help our environment . . . Why pollute when there is no reason to? . . . Why not recycle?" Pam is particularly concerned about toxins in the water, such as lead. Like many other interview participants, Pam is upset about the water crisis in Flint, Michigan. Her passion is evident when she explains her perspective on the issue: "When you are polluting and killing people" as is occurring "in Flint, Michigan . . . it is terrible." She particularly is affected by "these poor little kids that now have problems because of lead poisoning."

Pam's concern reflects widely available information on the water crisis in Flint. In 2014 the main water source for the city was changed from the Detroit water system to the Flint River. Following the change, local residents began to report problems with the water's smell and taste as well as physical symptoms indicating that the water was contaminated. People reported loss of hair, skin problems, unexplained miscarriages, and other negative health effects.[11] As Pam indicates, children are particularly vulnerable to the negative effects of lead-contaminated water. High levels of lead in blood can lead to brain damage, delayed development, social and emotional challenges, and trouble with hearing and speech.[12]

Pam, like many Americans, including several of the climate change skeptics we interviewed, is dumbfounded by the lack of governmental action taken in Flint and believes that "someone should go in there and immediately clean that up . . . this is harming people and instead we are taking away regulations." Flint has since switched back to its original water source and the state of Michigan has declared the water is now safe for consumption. Yet eight years after the contamination was first discovered, Flint continues to be plagued by water quality problems. Though the tested water sources indicate lower levels of lead, many residents continue to have corroded lead pipes transporting water to their homes.[13]

Flint is not alone in its problems with water quality. In fact, at the same time that the Flint water crisis was unfolding, approximately 6 percent of Americans (twenty-one million people) were relying on water sources not in compliance with the Safe Drinking Water Act for their daily consumption.[14] Perhaps reflecting that reality, 46 percent of the skeptics we surveyed were worried that "what has happened in Flint, Michigan, could happen where I live." That both air quality and water quality remain concerns for climate change skeptics is not surprising, given the persistent and growing problems the United States faces in these areas.

In addition to air and water pollution, the skeptics that we talked with also express concerns about soil health. One participant deeply passionate about this issue is Mark. Mark's most salient identity may be his political alignment. During our interview he repeatedly told us that he is a "Jeffersonian anarchist" who believes in very limited government and free markets. Mark believes that climate change is a hoax orchestrated by the United Nations and Al Gore, for personal profit. When we asked him whether or not people should be concerned about pollution, he responded "absolutely" and proceeded to an intensive discussion about "Big Ag," arguing that the pollution it produces "is horrible." Mark is particularly worried about "what it does to the soils and the oceans" and cites the "dead zone in the Gulf of Mexico" that results from "all the pesticides." In addition, Mark is worried about the impact pharmaceuticals have in "destroying the soil." He explains that "people poop or pee it out and there it goes." He attributes these pollutants to the loss of "pollinators," saying "there are reports from all over that there's not as many butterflies, monarchs, or bees." He expresses further concerns about human fertility: "I guess the sperm count of human males worldwide has been dropping drastically."

It is worth noting that in Idaho, where our interviews took place, agriculture is the single most significant contributor to the state's economy, making up 20 percent of its annual gross domestic product (GDP).[15] Many agricultural crops have historically led to erosion and poor soil health, but recent efforts to farm in more sustainable ways has gained traction in the state.[16] Though not a unique or significant problem in Idaho, there is evidence to suggest, as Mark does, that pharmaceuticals found in the water supply do end up in the soil via irrigation.[17] The impact of this on human health is yet unknown, but trace pharmaceuticals in the water supply has been shown to harm aquatic life.[18]

In addition to air, water, and soil contamination, skeptics express worry about the volume of trash humans produce and, specifically, the sheer volume of plastic waste. One person passionate about this issue is Logan. Logan is originally from a small farming town in the Midwest but he moved to Idaho to work in manufacturing. Logan believes that climate change is a natural phenomenon and that humans have no impact on the climate. When we asked him about pollution, he immediately mentioned "the plastic gyre in all the major oceans." Logan believes that plastic waste is going to "cause the oceans to change. That's going to increase the toxicity of the oceans." Thinking about future populations, he argues that "for us to keep the planet livable in the foreseeable future for your kids and grandkids, pollution is a problem." Beyond plastic in the oceans, Logan is also concerned about trash more broadly. He argues that human waste can affect water systems and have negative effects on human health.

Nearly 70 percent of the skeptics we surveyed express concern about plastic in the oceans. This topic has received increased news attention in recent years, as large islands of plastic have been found, including the Great Pacific Garbage Patch, a collection of over 1.8 trillion individual pieces of plastic estimated to weigh approximately eighty thousand tons, floating in the north Pacific Ocean.[19]

Participants such as David, Pam, Mark, and Logan express concerns shared by many of their skeptic peers. That they are concerned about issues that have gained media and activist attention in recent years demonstrates that skeptics are plugged in to current events and willing to engage intellectually and emotionally with environmental news stories.

We were surprised to find that nearly all of our interview participants have concerns about pollution. We also were puzzled as to how the skeptics we talked with could be concerned about pollution and not the associated phenomenon of climate change. The answer is deceptively simple: they do not see the two as connected.

For example, though Mark's primary concern is soil health, he also is concerned about air pollution. Yet, he argues, "carbon dioxide isn't a pollutant in my mind, and that's the one they've been touting regularly." He argues that scientists that suggest carbon dioxide is a pollutant have "ulterior motives." Mark believes that other chemicals in the air are the true toxins that pose a danger to humans. However, these pollutants are not alleged contributors to climate change.

Logan agrees. He further contends that "carbon dioxide is the current whipping boy on climate change" but that "carbon dioxide is 0.04 percent of our atmosphere." He then concludes that any change is "miniscule" and unlikely to impact the climate. Logan, like Mark, turns to other chemicals he feels are greater contributors to pollution, such as mercury.

Despite Mark's and Logan's contentions, carbon dioxide is a significant greenhouse gas. While the amount of CO_2 exhaled by humans is not significant enough to contribute to climate change, the volume produced through the use of fossil fuels combined with the reduction of carbon sequestration that occurs via deforestation, has led to a dangerous increase in CO_2 concentration in Earth's atmosphere.[20] However, the argument put forth by Mark and Logan, that CO_2 helps plants, is in fact accurate. But the increased ability for plants to photosynthesize as a result of increased levels of CO_2 are outweighed by the negative impacts of climate change on plant survival, including warming temperatures and changing water supplies (e.g., flood and drought).[21]

While nearly every person we interviewed said that they were concerned about pollution and that it should be a priority for the nation, two participants did not feel that pollution is an issue of warrant for the United States. When asked if they thought pollution is something Americans should care about, both pivoted and focused the discussion on pollution in Asia. The first, Ron, grew up in California but now lives in southern Idaho because, as he says, in California there were "too many people, too many idiots." Ron believes that climate change is a natural occurrence and that we are being lied to by scientists who are "bought out" by leftist politicians. When asked if pollution is a problem, Ron says he does not think it is a problem, but then goes on to criticize "China and Korea . . . That just blow pollution in the air like there's no tomorrow." He claims that people in these and other Asian countries cannot drink their water, and that they are "actually the biggest source of that pollution and everything . . . they don't even recycle." Ron believes that Asian nations want the United States to "take care of everything."

Ron is joined in his views by Ben, who also focuses his critique on Asia. Ben is originally from what he describes as a racially and culturally diverse town in Washington, but later moved to northern Idaho. He identifies as a political moderate because he is very concerned about racism in the criminal justice system. "We've gone so far and, it's like, you guys are police and racist," but he cares a great deal about his personal right to own a firearm, a belief which, he says, prevents him from being a Democrat. Ben believes that the climate is always changing and that anything scientists see as climate change is part of natural patterns and "less severe than we think it is." In talking about pollution, Ben also turns to Asia and uses personal anecdotes at first, arguing that "it's just disgusting, it's terrible there." He goes on to point out that the United States has better environmental regulations and that, "as far as pollution goes . . . it should be better controlled in other countries."

Ron and Ben are correct about the high levels of air and water pollution in China, which is primarily due to a lack of regulations on polluting companies.[22] However, when asked if people should care about pollution, the way that both Ron and Ben pivot to pointing blame at another nation—China specifically—reflects the historical xenophobic attitudes of the political right toward Asians. Among other stereotypes, American discourse surrounding Asians (and other immigrants) historically (and presently) has cast this group as "bring[ing] 'filth', 'immorality,' 'diseases.'"[23] In fact, the Chinese Exclusion Act, passed in the nineteenth century, the deportation of Filipinos in the twentieth century, and other anti-Asian legislation are due in part because of socially constructed arguments that Asians are "dirty."[24] At the start of the COVID-19 pandemic this discourse was seized on by then-president Trump, through the language he used to describe the virus that causes COVID-19 as the "China virus" or the disease as the "kung flu."[25]

Returning to the specific environmental concerns of skeptics, we see that climate change skeptics are worried about issues that receive significant media and activist attention. They are worried about air and water quality. They are concerned about soil health and plastic waste. In short: skeptics care about pollution. Perhaps this is something that Trump was aware of, and directly capitalized on, given his statements about clean air and water despite his anti-environmental and anti-climate policies. Among our participants, skeptics see pollution and climate change as unrelated. It is their contention that climate scientists view carbon dioxide as the primary contributor to climate change but that carbon dioxide is not a pollutant.

HABITAT DESTRUCTION AND SPECIES LOSS

In addition to air pollution, water and soil quality, and waste, skeptics are also concerned about the loss of animal habitat and species. In fact, this is the second most cited worry expressed by the skeptics we interviewed, with 58.6 percent saying they are concerned about habitat destruction and 59 percent expressing concern about deforestation. Approximately 55 percent are worried about animal species loss, while 63.4 percent fear bee population loss specifically, like Mark, mentioned earlier.[26]

These worries are not without warrant. Between 2001 and 2014 over 170 species went extinct worldwide in what many are calling the "sixth mass extinction."[27] Mass extinction events are defined as times within which over 75 percent of the species on Earth go extinct. The time boundaries for these events can vary but are typically on the order of thousands of years. Scientists believe previous mass extinctions have been caused by expansive volcanic activity, asteroids, and "a perfect store of multiple calamities."[28] Unlike previous mass extinctions, the current one is, undoubtedly, caused by human impact on the climate and the environment.[29] Currently scientists estimate that over 75 percent of all animal species are at risk of extinction because of human actions.[30]

Here we introduce two participants, Allen and Savannah, both of whom are particularly concerned about animals. Their comments summarize concerns expressed by a majority of skeptics in our sample regarding habitat destruction and species loss.

Allen grew up in a small conservative town in southeastern Idaho. While not a member of the Church of Jesus Christ of Latter-day Saints (LDS), most of the people in his town were LDS. He is apolitical because he finds politicians to be "all a bunch of dicks, really." Allen believes that climate change is occurring but thinks it is part of a natural ebb and flow. Yet he is very concerned about pollution, habitat loss, and animal species extinction. He explains this by saying, "even if climate change isn't real, there's still all the pollution and all the garbage in the ocean . . . It's affecting all the animals in the ocean and outside the ocean and things like that. So, yeah, even if it's not affecting the natural ozone layer it is still affecting animals."

Like Allen, Savannah grew up in a small conservative town in the southern part of Idaho. As an adult she moved north and now lives in the panhandle. Savannah got married when she was sixteen and perceives that event as the catalyst for her political awakening. She identifies as conservative in most areas and believes that "liberals have their heads up in the clouds a little bit." However, she finds solidarity with progressives on environmental issues. Savannah's environmentalism developed when she took a trip to the Oregon coast as a young teen, which launched a lifetime of care and passion for animals. Though Savannah believes that climate change is a natural cycle not caused by humans, she does have great concern for human impact on animal species. Her thoughts on this echo Allen's. Savannah explains that climate change is not a threat: "I don't really feel that Mother Nature's in any danger of being screwed up." But she does worry extensively about animal populations, saying, "There's danger in losing certain species of animals, which would be a tragedy in my mind."

Both Allen and Savannah reject that climate change is a problem but are concerned about animal species loss. As with air pollution, skeptics do not see these two issues as related. Allen, for example, blames plastic waste as causing animal species loss, rather than climate change. Other skeptics point to pollution and habitat destruction, but not climate change, as the causes of extinctions. Allen, Savannah, and their peers are not wrong. Animal species loss is caused by habitat destruction, pollution, and human consumptive habits.[31] However, scientists are also clear that climate change further contributes to and hastens extinctions.[32]

In their concern about animal species loss, Allen and Savannah both point to a level of environmental stewardship that many skeptics also express. For some, this sense of responsibility stems from their religious beliefs. For others, it comes from their own personal experiences with nature or negative environmental events they have observed. Allen expresses this sense of responsibility saying, "Well, like I said at the beginning of the interview, [pollution] really affects animals. We're not the only people out here or the only things on Earth. We have to share it with animals and we don't want to, well, we shouldn't destroy their homes too."

Our survey results show further support for animal habitat preservation initiatives among skeptics. Nearly 63 percent of participants support forest preservation initiatives and more than 70 percent support the preservation of national parks, while 57.6 percent support expanding the current US national parks system. These numbers are lower than among Americans more generally, 95 percent of whom want to secure the continuation of the park system in the future.[33]

Climate skeptics' concerns about animal species loss and habitat destruction are not wholly disconnected from their worries about pollution and waste. Allen and Savannah both mention the impact of pollution on animal habitats. Mark, too, is very concerned about the loss of pollinators as a result of toxins. As with the issue of pollution, the causes of animal species loss are connected to climate change and this shared care between skeptics and those who accept climate science presents possibilities for widely supported pro-climate policy initiatives.

RENEWABLE ENERGY

The concerns that skeptics demonstrate about pollution and habitat loss may influence what arose in our interviews as the third most significant issue: energy. Despite Trump's moves during his presidency to expand offshore oil drilling, including in the Arctic National Wildlife Refuge, and his efforts to reduce incentives for renewables and cleaner energy, the skeptics we interviewed are interested in alternative energy systems.

Many participants support research into both solar and hydroelectric energy.[34] Among respondents, over half support investment in solar farms and 49 percent approve of expanding hydroelectric dams. Consider, for example, Bill's perceptions. Bill grew up in eastern Washington and moved to Idaho in his early adulthood. He lives in a rural community and works at a local big-box store. As an adult Bill converted to the Church of Jesus Christ of Latter-day Saints and remains an active member. He identifies as politically conservative and believes that climate change is occurring, but is not caused by humans. Regarding energy, Bill contends that "exploring alternate energy sources is a great idea." He particularly "want[s] to see more" exploration of solar and hydroelectric energy. He, like many of his peers, wants to "get away from more of these energy generators, such as coal, that pollute the environment, and go into more sustainable ones." Yet he worries that until these systems are more affordable, "we're kind of stuck with what we have."

As Bill suggests, coal is a significant contributor to air pollution. When burned, coal produces several harmful pollutants, including mercury, sulfur dioxide, nitrogen oxides, and other chemicals. All of these toxins can cause problems with breathing and the circulatory and nervous systems, among other health impacts.[35] Coal also negatively impacts water quality and is a key contributor to climate change.[36] Coal provides approximately

30 percent of the energy used in the United States.[37] Fewer than half of the climate change skeptics support increasing the use of fossil fuels: only 33.5 percent support expanding fracking and only 38.6 percent support extending offshore oil drilling.[38]

Skeptics often mention the issue of cost and affordability when discussing energy. One participant who explains this well is Karen. Karen grew up in Utah but for much of her life has lived in southern Idaho. She lived in isolation from others as a child, except for connections to her LDS community. Karen identifies as a Republican but would be willing to vote for the "right" Democrat. On the issue of energy, Karen expresses a frustration that she shares with many of her skeptic peers. Specifically, that alternative energy sources, such as electric vehicles and solar panels, are cost prohibitive. She contends, "They're so expensive. They're so unobtainable. Most families can't afford a smart car that runs on batteries or solar." Karen wants to "feel like [she's] doing some good" and has a "community mindset," but she feels it is simply cost prohibitive. Our survey respondents indicate some support for either tax breaks for the production of energy efficient vehicles (45.9 percent support this measure) or government regulations for fuel efficiency standards in new cars (41.6 percent).[39]

Concerns about the cost of alternative energy sources is valid. The average residential solar installation in Idaho costs over $11,000 after the federal solar tax credit, while the average cost for an electric car is nearly $56,000.[40] Under the Obama administration a federal tax credit was passed for the purchase of both of these items, though under both Obama's and Trump's administrations that credit was annually decreased.[41] In 2022 the Biden administration proposed a novel tax credit for American-made electric vehicles of up to $12,500 and a reduction in rooftop solar costs as part of their Build Back Better initiative, currently awaiting Senate approval.[42] In contrast, the United States spends over $649 billion on oil subsidies annually, which significantly reduces the cost of this fossil fuel for consumers.[43] Despite a lack of similar investment, the cost of renewable energy systems has been decreasing. Solar energy costs, for example, have decreased 73 percent since 2010.[44]

Beyond mitigating pollution, the people we interviewed also see renewable energy systems as key to what conservatives often call "energy independence." For example, Jake is concerned about the move to make more electric cars without concurrently finding energy alternatives to coal. He argues that "we don't have a solution" for the increased demand in electricity that will result. In general, people who seek energy independence are concerned about the United States' reliance on oil from the Middle East.[45] Conservatives commonly believe this will economically benefit the United States by expanding the use of American fossil fuel sources.[46]

It is also worth noting that wind energy receives less support than solar energy among some of our interview participants. Many participants note that wind energy is not cost effective and express puzzlement that, when they drive past turbines, they often are not spinning. One of the exceptions is Jake, who finds all alternative energy development exciting. He glowingly remarked, "We need to focus on finding alternative sources of energy. The solar project that's going on out there in the desert on the way to Mountain Home, the wind farms. The more of that that I see the happier I am about it. I just wish that there was more funding going into it instead of SUVs."

More than half of the climate change skeptics we surveyed and interviewed support investment in renewable energy. Many also are in favor of regulating the fossil fuel industry. As with the issues of pollution and animal species loss, these skeptics do not connect their support of these areas with support for climate change. Rather, they see an investment in alternative energy systems as advantageous in reducing pollution and moving the United States toward energy independence. Over the past decades the use of renewable energy systems has increased but so too has the use of natural gas. Coal, on the other hand, has decreased in its share of the US energy portfolio since approximately 2000.[47] As with pollution and habitat destruction and animal species loss, energy is a third area ripe for cross-ideological cooperation and policy building.

OTHER PRO-ENVIRONMENTAL ATTITUDES AND BEHAVIORS

The top three environmental issues that concern climate change skeptics are not the only environmental topics about which they care. Some participants are worried about the funding cuts the Trump administration made to the EPA, and many are concerned about environmental events like forest fires. What is most remarkable, though, is how strongly the people we interviewed feel a sense of responsibility to curb pollution and waste, to protect animal habitats, and to, as Karen shared, "feel like [they're] doing some good." In fact, we find that skeptics behave in a number of pro-environmental ways. Of the people we surveyed, 87 percent turn lights off when they don't need them, 86 percent recycle in the home, 80 percent monitor their heat to reduce energy use, and 73 percent claim to have replaced their incandescent lightbulbs with LED ones. While one could easily argue that these are behaviors that may economically benefit consumers and thus are not exclusively pro-environmental acts, 75 percent of skeptics self-identify as "the kind of person who makes efforts to conserve natural resources," indicating that such pro-environmental actions may reflect more than a simple concern for the bottom line.

When considering pro-environmental attitudes and behaviors of skeptics more deeply, we find that personal experience—both positive and negative—serves as an important stimulus for pro-environmentalism.[48] Our survey data shows that, among skeptics, men who are politically conservative, people with less education, and those who are distrustful of science tend to be less pro-environmental than other skeptics. Furthermore, skeptics who believe that climate change is a hoax and skeptics who believe climate change is ordained by God also harbor less environmentalism.[49] We now turn to a discussion of the association between personal experience and environmental concern, which emerged in our research as pertinent to understanding climate change skeptics' environmental attitudes and behaviors.

Personal Experiences

Savannah discovered a passion for animals as a teenager, during a vacation to the Pacific coast, a positive environmental experience that served as a catalyst for her broader dedication to this issue. Similarly, Sam, who believes that climate change is a natural phenomenon, also feels a sense of responsibility and stewardship for the environment as a result of his childhood experiences. Sam grew up in the Midwest and moved to Idaho as an adult. Expressing a sort of nostalgia, he recalls: "From a young age, my dad would just be going on walks. He would pick up beer cans and things from the road as he was walking along the ditch. He would just bring a bag with him and pick up and it would be great. He would collect them and then take them to the recycling." Now, as an adult himself, Sam identifies as "a big outdoorsman." He hunts and fishes and "love[s] to be out in nature." He finds pollution, trash, and evidence of human interference in wilderness areas "jarring." He worries about human impact on air and water systems and hopes to motivate others to care for the wilderness.

Personal experiences with the environment, such as those described by Sam and Savannah, are consistently documented as influencing a level of environmentalism later in life. For example, childhood experiences such as spending time outdoors, talking about environmental issues, and watching nature shows increase environmentalism in adulthood.[50]

For some skeptics, the personal experiences that led to their environmentalism were largely positive. Yet for others the experience was a negative environmental event that catalyzed their care. Pam, for example, had a family member that worked in a vermiculite mine. After years of service he died from occupational exposure to asbestos. Pam has other family members who are also sick from working in the same mine. She recalls that children in the community were exposed to asbestos on playgrounds and that many people are very sick. She credits the EPA for coming in and "digging everything up" and concludes, "We have to help people."

While several of our participants had negative experiences with toxins as a result of the mining industry, others experienced air pollution in their towns. Jennifer, for example, jokes about how Nampa, a city in southern Idaho, always "smelled like peanut butter and onions," but then she resolutely recalls that "it does affect your life." Lewiston, a town in northern Idaho famous for its paper mill smell, was also repeatedly identified as evidence of personal experience with low air quality. Zeke explains well the connection between air pollution and health and well-being. He recalls that his "first wife had asthma and when the air quality in the Boise Valley got bad in the winter she had trouble breathing." As a result, they moved to northern Idaho and her "quality of life" improved. More recently, Zeke's girlfriend "had some major heart issues and during the winter she had a lot of breathing issues because of that inversion." Zeke emphasizes: "I can tell you pollution in the air is a problem . . . the amount of carbon and other garbage that we're pumping into the air, we can do something about that, but I don't know what."

That a negative environmental event has led Pam, Jennifer, and Zeke to greater level of pro-environmentalism is not altogether surprising. Social science has long documented that personal experiences influence personal actions and values regarding the environment. For example, people who live near toxic waste sites are more likely to hold pro-environmental views than others.[51] Further, for the general population, exposure to "problem sites" such as "landfills or waste disposal site[s]" correlate with pro-environmental beliefs and behaviors.[52] Finally, exposure to environmental disasters (e.g., the Fukushima, Japan, nuclear accident) leads to increasing public risk perceptions, pro-environmental attitudes, and behavioral intentions.[53] While most people who perceive danger do work to reduce risk, if the consequences of environmental disasters experienced are not severe, they may be less inclined to engage in mitigating behavior.[54] Overall it is worth considering that the strong association between personal experience and pro-environmentalism suggests that when someone experiences a negative environmental event that is more clearly connected to climate change (such as severe flooding, wildfire, an intense hurricane, or visible sea level rise) they may become less skeptical of climate change.

Any discussion of personal experience with nature—whether positive or negative—and its impact on skeptics' environmentalism echoes certain emotions. Sam appears nostalgic. Savannah is worried. Pam is afraid. Jennifer and Zeke both come off as disgusted, suggesting that there may be an emotional component to environmentalism for skeptics rooted in the lasting emotional effects of a negative or positive environmental experience. Research on emotions and pro-environmental behavior suggests that there is a positive correlation between certain emotions and eco-friendly attitudes and actions (explored further in chapter 6). Guilt and nostalgia appear to motivate pro-environmentalism.[55] Sadness, too, is motivating, but only in the short term.[56] Fear is often motivating with regards to short-term actions but not for long-term behavioral changes; worry, a longer-acting and less intense version of fear, may motivate change in distinct ways.[57] In this regard, skeptics behave, at least environmentally, in ways parallel to the general population.

CONCLUSIONS AND IMPLICATIONS

The people that we interviewed reject anthropogenic climate change. While some believe that climate change is simply not occurring, others, like Donald Trump, argue that it is a natural occurrence and not caused by human activity. Yet skeptics share Trump's expressed concern about air and water quality. Pollution and associated waste are the primary concern expressed by climate change skeptics. They also care about habitat and animal species loss as well as US energy systems.

The pro-environmentalism we find among skeptics and the factors that contribute to the strength of these sentiments have significant implications for both climate communication and climate policy and provide hopeful avenues for collaboration across an ideological chasm. First, personal environmental experience could be channeled to advance climate communication. Furthermore, our research suggests that evoking certain emotions such as worry and nostalgia can lead to increased support for certain pro-environmental policies.[58] We suggest policy initiatives that both align with the environmental goals of skeptics and serve to mitigate climate change.

While prior research on risk communication indicates that both direct experience and indirect experience through education and media can affect perceptions and behavior related to environmental risks (for example, people can empathize with stories of other people suffering following a natural disaster), as far as climate change skeptics' pro-environmental views are concerned, direct experience seems to hold a higher importance.[59] As such, climate communication methods that evoke direct, hands-on experience is best-suited for environmental campaigns that target skeptics.

Targeted climate communication strategies could draw from negative environmental events that are more clearly connected to climate change (e.g., certain natural hazards). For instance, Gisela Wachinger and her co-authors recommend experiencing "the power of water by walking through a river, listening to lively narratives from hazard witnesses, studying historical accounts of past disasters (e.g., flood marks in medieval European cities or shrines in Japan)" as a way to heighten public risk perceptions of environmental disasters and associated personal preparedness actions.[60]

Where actual physical experiences cannot be evoked, new technological platforms such as virtual reality (VR) can provide powerful simulations of environmental experiences, both positive and negative. The Virtual Human Interaction Lab of Stanford University follows in this line of research, examining how VR experiences affect pro-environmentalism.[61] When investigating the effects of direct and indirect experiences, both real-life and simulated, future research might compare these effects across climate change skeptics and nonskeptics.

Knowing that skeptics' environmentalism evokes emotions such as worry, nostalgia, and fear, suggests that feelings associated with environmental experiences are important and likely can be recalled via climate communications. For example, when climate communication efforts are connected to certain "objects of care," such as a specific animal species or a habitat, the psychological distance between climate change and the individuals themselves is reduced, likely increasing environmental concern among skeptics.[62]

Savannah, for example, is very concerned that her children will not "be able to experience the animals and the different fascinating life forms on this earth." This statement reflects both a sense of responsibility for animals but also for future human generations. It further evokes a feeling of nostalgia. This combination of feelings may motivate action in a way that fear-based messaging would not. If new technological platforms such as VR prove effective in evoking emotional reactions to specific environmental experiences and "objects of care," as early research indicates, these mechanisms may provide more avenues to garner support for pro-environmental policy initiatives among skeptics.

As far as climate-related environmental hazards are concerned, the most severely affected communities in the United Stated are low-income populations of color, constituting a portion of the American public that is less likely to reject the physical realities and human causes of climate change.[63] Yet, as the intensity and frequency of climate change–induced natural disasters increase, no community will remain completely unaffected. The wildfires in the American West over the past few years, for example, have engulfed homes and properties of individuals across socioeconomic lines, damaging properties of regular residents, public officials, and wealthy celebrities alike.[64] These experiences, both direct and indirect (associated with extensive media coverage), may act as powerful indicators, for both skeptics and nonskeptics, that climate change no longer only affects "geographically and temporally distant people and places," potentially heightening skeptics' pro-environmentalism and support for climate change mitigation strategies.[65]

The level of care that skeptics have for environmental issues also has important implications for both environmental and climate regulations and policies. In fact, focusing on air pollution alone—the number one environmental concern for skeptics—can have profound impacts on climate change. The United Nations Environment Programme (UNEP), for example, suggests that "immediate changes to air pollution levels also have immediate effects" on climate change. Specifically, a reduction in "short-lived climate pollutants" such as "methane, tropospheric ozone, hydrofluorocarbons and black carbon" will reduce "the chances of triggering dangerous climate tipping points."[66] When considering carbon dioxide, a longer-lasting pollutant, a reduction in emissions to improve air quality and save human lives would significantly impact climate change as well.[67]

Similarly, policies that preserve animal habitats would also positively impact climate change. In fact, deforestation is one of the leading contributors to climate change. The act of cutting forests is a significant greenhouse gas contributor but, further, forests are needed to sequester carbon dioxide in the atmosphere.[68] This is also true of efforts aimed at expanding the use of renewable energy. In the United States about a third of carbon emissions come from electricity use, primarily coal and natural gas manufacturing.[69] The US Department of Energy's National Renewable Energy Laboratory claims that it is possible to move 80 percent of US energy production to renewable sources by 2050. Doing so, the lab estimates, would lower emissions from electricity by over 80 percent.[70]

We were surprised to find that skeptics support federal regulations to achieve their environmental goals, even though skeptics also recognize that such policies run counter to many of their libertarian political tendencies. Even though the political right is typically opposed to government regulations, climate change skeptics largely support federal regulations for specific environmental protections. A few dissenting voices expressed a preference for state-mandated policies instead.

Skeptics simply believe that the environment is too important and needs to have regulatory protections for its preservation. For example, Pam connects the environment to the "safety and health of our country" and, on these issues, believes that "regulations that keep companies and people from polluting rivers and polluting the air" should be passed by the federal government. She believes such policies "should be enforced big-time" and the "larger the company, the more regulations they need."

Pam also makes a second argument, that "homeowners like me don't understand all of the things that they could be doing." This secondary point was voiced by many skeptics. Jill, for example, contends that people are "just unaware of what's really going on" and believes that the government "has to try what they can to keep people from swaying completely and not caring about pollution or sustainability."

A few skeptics disagreed with the consensus that regulations are needed for environmental protection. One such skeptic is David. David is a Libertarian who believes that "free market solutions to air quality and water quality" are ideal. He believes that consumers will not make purchases from companies that pollute and that "free market competition" will serve as a proper "incentive" for companies.

The handful of skeptics who do not support regulations at the federal level do seem to agree with the role and importance of regulations at the state level. For example, Lee argues there is a role for regulations, and that they are "necessary." However, he believes "it should be left up to the states . . . because of the different environments." He contends that, in California, "it's probably not a bad thing that they put extra taxes on you if you drive by yourself . . . because Los Angeles has some of the worst air pollution in the world, and that's not good for anybody." But, he argues, these same policies don't make sense in Idaho, "where it is really green." Lee believes that in Idaho "there should be more freedom to use vehicles."

When designing communications strategies to garner skeptics' support for climate change mitigation, it is important to note that the language used to discuss and frame environmental issues is particularly important. Many skeptics care about preserving public lands for hunting or believe in "energy independence" and therefore support investment in renewable energy systems. The terms that are used to engage skeptics, or conservatives more broadly, can significantly impact their willingness to hear and adhere to pro-environmental messaging.

We are not attempting to tout climate change skeptics as brazen environmentalists. In fact, on this measure their attitudes and actions pale in comparison to Americans in general and especially compared to political progressives. Rather, we note that there is a surprising level of environmental care and concern among skeptics that could be used to garner support for policies that will mitigate climate change. One skeptic, Jodie, summarized this well: she approves many policies enacted to "combat climate change." Though she doesn't believe that climate change is a concern, she recognizes that these policies "are good for the earth anyways" and thereby worth employing.

five

ENGAGEMENT WITH MEDIA AND INFORMATION

"COVER THE NEWS. TELL THE STORY. THAT'S IT." With these words Greg summarized his views on media coverage of climate change and ended his interview. Greg, a white man with a postgraduate degree, grew up in the San Francisco Bay Area in an "average middle-class community." Greg recalls becoming politically aware during the 1976 "Jimmy Carter election," which "was like a horse race or a football game where you root for your team." Later, Greg and his family moved to Idaho because it's a "good place to raise a family." Eventually they settled down in the Boise area. Greg posits that his climate change views are "nuanced." To him it is plausible that climate change is just a natural phenomenon. As such, he questions the undue authority granted to scientists "as high priests of knowledge that tell us things that we're supposed to say."

When we introduced Greg we noted how he harbors significant doubts about climate science, including what he perceives as a skewing of scientific data to support various hidden incentive structures and career pressures. When further examining Greg's views on climate change, we see that his distrust and skepticism of climate science extends to and stems from his perceptions of the media. Greg receives his information, including information about climate change, from "a lot of places," spanning across the political spectrum,

> If I'm looking for information, I'll probably look at the *Wall Street Journal*. I would say, their bias is center-right. I would say the *New York Times* is biased center-left, maybe a little lefter. *Washington Post*, center-left. And I think if you understand the bias, then you go, uh, I don't know that they're making stuff up, but you can read different coverage of same events and learn quite a bit. I don't watch TV news because it's just a mess. I can't deal with it. Everything's breaking news and I just can't deal with it.

Further elaborating his views on media bias, Greg explicates how he combats participating in an echo chamber. He purports to use peer-reviewed academic articles and technical reports to verify information reported in the mainstream media. The list of primary sources he named during our interview included the IPCC reports, JSTOR database (a digital library of academic publications), and Marginal Revolution (a popular right-leaning blog spearheaded by American economists Tyler Cowen and Alex Tabarrok).

Greg contends that exposing oneself to multiple sources and perspectives and verifying information reported in the media by cross-checking is essential, because "journalists have their limitations . . . no journalist is an expert in everything they cover. Very few journalists are experts in anything they cover." Further, Greg states, it is important to understand the inherent biases in the media so as to "be aware of that."

While Greg believes media does a "decent job" overall on reporting about climate change, he adds a caveat: "They're trying to sell papers so they may make more of something than it is." In his view, biased, inaccurate, and sensational reporting occurs due to explicit and implicit political agendas but also other, often perverse incentives. "They're looking for conflict and you got to find the fight here so they may play that up when it doesn't necessarily exist. There may be a lot more consensus than you think."

Information and its framing as obtained through various media and information sources are core to development of an understanding of climate change and an individual's role within this politically charged issue. Lacking first-hand knowledge about climate change, skeptics, much like the broader public, must rely on experts, such as scientists, journalists, and other opinion leaders as sources of information when forming their beliefs about the topic.[1] It is true that skeptics develop their identities in opposition to climate scientists. However, this identity formation does not occur in a vacuum. In fact, media and information portrayal are a significant cultural force that shapes public opinion on complex topics, including climate change. As such, media and information drives skeptics' beliefs about climate change, including their opposition to and distrust of climate science and scientists.

Thus we begin this chapter with an examination of prior work on the media and information landscape within which skeptics form their beliefs, paying close attention to misinformation and disinformation that emanates from the climate change denial countermovement. We then explore the sources from which skeptics access their information, their levels of trust and distrust of specific information sources, how they engage with this information, and the types of information skeptics employ in the creation of their own stories about climate change. We conclude by demonstrating how engagement with certain media and information may deepen the skeptic identity but may also provide an opportunity for better climate communication.

MEDIA EFFECTS ON OPINION FORMATION

While there is overwhelming scientific consensus on the existence of anthropogenic climate change, public opinion about it is not a direct reflection of this consensus nor of the accuracy of scientific information received. Quite the opposite: research conducted by prominent environmental sociologist Robert Brulle and his colleagues suggests that, in fact, scientific information has little effect on public opinion when compared to political commentary, a much more significant contributor to opinion formation on controversial issues.[2] In particular, elite partisan battles and conflicting priorities (environmental concerns versus economic concerns, for example) that are typically conveyed to the public via media, along with the overall quantity and quality of media coverage, shapes opinions the most.[3] On climate change, decades-long social science analyses suggests that media and information, both traditional journalism and explicit disinformation campaigns waged by the denial countermovement, have significantly affected perceptions.[4]

Some media coverage of climate change contributes to the elevation of the "outlier voices" of skeptics, contrarians, and denialists, who tend to garner disproportionate visibility. Inflated visibility enables outlier voices to leave significant imprints on the public debate surrounding climate change.[5] Media scholars argue that this happens because mass media is faced with two key challenges when reporting about climate change, both of which stem from common journalistic norms, values, and practices.

First, mass media "conflates the messenger" by treating all expressions of skepticism alike and failing to distinguish skepticism derived from ideology from skepticism derived from evidence-based science. Second, mass media "conflates the message" by failing to situate climate change coverage within the context of the larger, contemporary scientific view.[6] Despite the scientific consensus, news outlets also persist in what Maxwell Boykoff and Jules Boykoff call "balance as bias." In an attempt to present a balance of opinion, they overrepresent pieces of climate change skepticism which thereby inflates uncertainty about the issue. In essence, "adherence to the norm of balanced reporting leads to informationally biased coverage of global warming."[7] In other words, minor instances of scientific disagreement and uncertainty get disproportionate media coverage. Partly as a result, roughly about one third of articles in US print coverage present some form of a skeptical argument when presenting content that is related to climate change, despite overwhelming scientific consensus on the issue.[8]

In addition to bias in coverage, communication scholars have also noted the effects of framing on public opinion formation. Much like the informational content of news stories, the way those stories are told, the frames used by journalists, also provides important contexts that can influence how people interpret news stories and associated information.[9] Frames are an inherent part of the communication process. They present "why an issue might be a problem, who or what might be responsible for it, and what should be done about it."[10] Framing serves to pare down information and emphasize certain elements over others. The keen scholarly interest in framing is reflected by the fact that a simple Google Scholar search of the term "news framing" yielded 21,700 hits between January 2021 and April 2022.[11]

Communications scholar Matthew Nisbet has developed a typology of media frames applicable specifically to climate change coverage. He finds that the denial countermovement employs two key frames in their stories challenging climate science: the "scientific and technical uncertainty" frame and the "economic development and competitiveness" frame. These two frames convince viewers and readers that climate science is uncertain and unsettled, and that taking the actions required to mitigate the effects of climate change will lead to dire economic consequences.[12]

Because audience members filter media representations of information through their personal experiences, ideologies, and identities, framing does not affect everyone alike. Research by communication scholars Graham Dixon, Jay Hmielowski, and Yanni Ma shows that when it comes to US conservatives, framing news stories around scientific consensus may not be an effective technique, while messages that emphasize free market solutions to climate change show the potential to increase conservatives' acceptance of the phenomenon.[13]

Interesting nuances in media effects on climate change opinion have also been uncovered regarding partisan media coverage. Recent scholarship on information and media has pointed to the phenomenon of the "echo chamber," wherein people seek out information that confirms their preexisting beliefs. Within this echo chamber, partisan media reinforce the preexisting views of audience members who share the same partisan identity. When these audience members are presented with conflicting or divergent views elsewhere, they engage in "motivated reasoning" and confirmation bias, leading them to reject the contradicting information.[14] Media scholars have termed this phenomenon the "boomerang effect." Predictably, political ideology and party identification is the main source of motivated reasoning. As a result, audience members with strong political identities are more likely to reject coverage that is inconsistent with their views when the coverage is seen as politically biased.[15]

The Denial Countermovement

While the uncertainties and complexities of anthropogenic climate change make it challenging for nonexperts to grasp its causes, effects, and appropriate responses, it is not only this complexity that leads to a relatively high level of skepticism in the United States.[16] In fact, the "climate change denial countermovement," launched primarily by the US fossil fuel industry (most notably ExxonMobil and Peabody Coal) and its associated front groups, conservative foundations, and think tanks, is largely responsible for sowing doubt about climate change in the United States.[17] They do this mainly by employing conservative and right-wing media to disseminate messages that contradict the prevailing scientific consensus.

The denial countermovement has developed an "extensive coalition structure" among multiple organizations that have their own individual coalitions (e.g., Cooler Heads Coalition, Alliance for Energy and Economic Growth, Coalition for American Jobs), and some organizations participate in various coalitions at the same time. Together these coalitions primarily represent the interests of coal, rail, steel, and electric utility companies.[18] These groups continue to "manufacture uncertainty" about the realities and human causes of climate change by employing organized "disinformation campaigns" that engage in persistent attacks on climate science and scientists.[19] These countermovement actors are also supported by conservative politicians and a small number of contrarian scientists who refute climate science.

The conservative media, a powerful actor in the denial countermovement, contributes to consensus skepticism by claiming a lack of scientific consensus on anthropogenic climate change.[20] As a result, scholars find a connection between conservative media use and increasing uncertainty about climate change.[21] Because of the power of disinformation campaigns to generate skepticism, sociologist Riley Dunlap and his colleagues have labeled the actors within these campaigns as the "denial machine." Unsurprisingly, *Fox News* and conservative talk radio programs have been singled out by these scholars as "key elements of the denial machine."[22]

Recent research by Yale sociologist Justin Farrell, which employs novel computational data collection and analysis techniques, shows that "contrarian organizations" that receive corporate funding are more likely to disseminate content that polarizes attitudes toward climate change. This includes grassroots lobbying firms that work for larger corporations, as well as industry groups and associations.[23] Further complicating this information landscape, climate change misinformation is also propagated by US philanthropic organizations that are tied to corporate donors.[24] Overall, well-organized and well-funded "contrarian campaigns" spread skepticism of climate science, primarily through the development of an alternative "contrarian discourse" perpetuated via conservative media, according to Farrell.[25]

ACCESS TO MEDIA AND INFORMATION

These effects are reflected in the case of self-declared climate change skeptics. Consider that skeptics in our sample claimed to rely on a variety of sources for their climate change information. When we examine the extent to which they access different sources, we find that they primarily access only conservative media such as the Fox Network, ignoring mainstream or liberal media. This was illuminated when Jane, a white woman in her mid-sixties who identifies as "leaning Republican," stated, "The only news channel I watch is Fox." In our survey, the average degree of reliance on the Fox Network for climate change information expressed by skeptics (2.79 on a 1–7 scale, ranging from "not at all rely=1" to "rely a great deal=7") was significantly higher than that they expressed for CNN (2.21) and NPR (2.41). Right-leaning peripheral media outlets (e.g., One America News and Infowars) were also cited by skeptics, though less frequently than *Fox News*.[26]

From our interviews it appears that skeptics also rely on non-mainstream sources of information such as online news aggregate sites, blogs, podcasts, and social media. For example, consider the media habits of three of our interview participants, Bill, Nick, and Douglas. It is evident from their comments and those of numerous other skeptics that their climate change information comes largely from the conservative echo chamber, consisting of the Fox Network, conservative talk radio, skeptic blogs, and contrarians with a large social media presence. By existing within this echo chamber, skeptics affirm their preexisting views about climate change and thus strengthen their skeptic identity.

Bill, the Libertarian with a college degree, used to listen to NPR, "especially during the Bush years," but he feels that when Barack Obama was elected president, NPR did not provide critical coverage of the Obama administration. "Obama did a whole bunch of things that NPR and a lot of other radio media outlets just completely glossed over. And that was one of the things that sort of made me realize that a lot of the mainstream media really can't be trusted. I mean they tend to have an agenda as well. That's when I started looking into more conservative things." Bill doesn't "fully believe" the reporting of conservative media either, but he does rely on talk radio, and, at that time, listened frequently to Rush Limbaugh, Mike Gallagher, and Glenn Beck, all conservative political commentators and talk radio hosts: "For talk radio hosts, actually the one that I like the most so far has been Glenn Beck."

It is worth noting that Glenn Beck has been a climate change conspiracy adherent since the early 2000s. In fact, Beck has included global warming in a list of "destructive lies" that he perceives were deliberately developed to mislead the American public.[27]

Beyond talk radio, Bill has a Facebook account that he uses (infrequently) to access some groups' chat rooms and websites, but in his own words, "they are opinion sites, and I know they're opinion sites, and one of them is called Chicks on the Right." Bill has a positive view of Facebook—despite recognizing that what he sees on it is largely based on personal opinion—because he perceives it as a reliable aggregate of conservative viewpoints that is both transparent and fair:

> It [Facebook] is a great website too, especially if you're interested in looking at conservative media. And all of their stuff is opinion pieces but they tend to include links to the articles where they found the information, so they're willing to express their opinions about these things but you can go ahead and follow it to see if you agree with their thoughts. Now they're not just going to link to other conservative news sites. I mean they link to Huffington Post, CNN, things like that.

It appears that Bill largely consumes only conservative media, inadvertently subjecting himself to a talk radio and social media "filter bubble," which also likely affects the climate change information he is exposed to. This shapes the understanding of climate change that he develops.

Coined in 2011 by tech entrepreneur Eli Pariser, the concept "filter bubble" refers to recommendation and personalization algorithms used in social media and search engines that can arguably exacerbate divisions by isolating audience members in unhealthy and unbalanced information systems, thus further polarizing attitudes toward controversial topics.[28] As a result, people no longer just have differences of opinion when it comes to controversial issues; their opinions are often built on nonoverlapping factual bases.

Nick, the conservative Democrat, also relies mainly on conservative and right-wing media for his information and has a preference for online news aggregate channels. As he says,

> I like the Drudge Report and Zero Hedge. Zero Hedge has some great stuff on it. Drudge is ninety percent news and interesting articles and ten percent tabloid stuff, which sometimes is entertaining. I like lewrockwell.com, which is a compounding of different columnists and writers, and every day they'll have like a dozen. But then all the old stuff is there too, so if you find someone you really like you can go back and read everything they wrote. Infowars, Prison Planet, they do some of their own stuff but also aggregate. And have some really bright people on interviews. They're on YouTube too. Well, stuff that just pops up on YouTube. Anyways, they have algorithms or cookies or something, so after you've been to some sites, ones related will pop up on the home page or the column on the right. So I find really fascinating things that way. Interviews on health, nutrition, and science. Anything that's science, archaeology, astronomy, if I see that in the headline, I'll read it 'cause I find it interesting.

The Drudge Report is a conservative aggregation website; Zero Hedge is a far-right Libertarian financial blog; lewrockwell.com is a conservative leaning blog; Infowars and prisonplanet.com, both from Alex Jones, constitute far-right media oriented toward conspiracy theories. In fact, Infowars, which has been accused of peddling conspiracy theories, was permanently banned from several social media channels, including Facebook and YouTube, in 2018.[29]

The types of information Nick accesses and the ways he engages with this information (e.g., via YouTube recommendations), once again provide a clear case of someone steeped in a social media filter bubble. Existing within this bubble potentially exacerbates Nick's confirmation bias, leading him to accept claims that align with his preexisting beliefs and ignore dissenting claims. Research finds that confirmation bias, or the tendency to prefer messages that align with preexisting attitudes over messages that challenge them, is especially acute among online users because of algorithmic tendencies as well as the easy aggregation of like-minded people online.[30]

Claiming to access a range of sources, Douglas, the apolitical white man from northern Idaho, appears similar to Bill and Nick in that he relies mainly on conservative talk radio and commentators for his climate-related information. In his interview Douglas explains,

> One of my favorite guys actually is Michael Savage, who is pretty far Libertarian, pretty far right. He's a bizarre character because he is a Jew and a son of an immigrant, and he is a conservative and like a hardcore environmentalist at the same time. And very pro-science, so he's an interesting character. I listen to NPR also, I listen to, um, those are the main two news outlets I hear on the radio. Don't really watch any news on the TV or the internet. I'll read articles in different periodicals. For science stuff I'll pick up a copy of the *Smithsonian Magazine* or *Discover* magazine. I'll read some *Washington Post*, *New York Times* articles occasionally, as well as *Daily News* and *Tribune* articles.

Michael Savage, an American conservative author, political commentator, and host of the AM radio talk show "Savage Nation," is a well-known climate change denier who has once claimed "the greatest fraud in American history is the concept of man-made climate change."[31]

All in all, Bill, Nick, and Douglas exist within echo chambers that align with their ideological belief systems as well as their political and skeptic identities. While they list information sources from across the political spectrum, their stated preferences directly contradict their expressed need for accessing information from multiple, ideologically diverse sources. Instead, skeptics' climate change information seems to stem primarily from actors within the "denial machine." For Bill, this actor is Glenn Beck, a climate change conspiracist; for Nick, Infowars and other right-wing websites that peddle in conspiracy theories; and for Douglas, Michael Savage, a vocal climate change denier. It is likely that these information sources strengthen the skeptic identities of Bill, Nick, and Douglas, who, respectively, describe their overall climate change views as "I am extremely skeptical," "I full-out dissent," and "We don't know enough."

Research shows that actors within this echo chamber, particularly conservative columnists and bloggers, are especially problematic because of their popularity among skeptics. When skeptics contribute to the wide circulation of opinion columns, these actions sow further doubt about climate change. In fact, when Shaun Elsasser and Riley Dunlap analyzed over two hundred op-eds written by conservative columnists that they gathered from perusing self-declared conservative websites, they found conservative op-eds to be "highly dismissive" of climate change and ripe with common skeptical arguments such as "It's not happening," "It's not us," "It's not bad," and "It's too hard [to solve]."[32] More recently, researchers have noted that political elites' opinions on whether climate change is anthropogenic or not has become "the central organizing force behind echo chambers in the US climate policy network."[33]

TRUST IN MEDIA AND OTHER SOURCES

Earlier we demonstrated that skeptics overwhelmingly rely on conservative media for their information on climate change. Concurrently, our survey data reveals that they lack general trust in mainstream media. Only 33.4 percent of skeptics in our sample expressed any trust in mainstream media, compared to 54 percent who trust Republican leaders, 54.2 percent who trust the Trump administration, and 64.5 percent who trust private businesses.

It is not entirely surprising that respondents express greater trust in Republican leaders than in Democratic leaders (54.2 versus 30.5 percent); this reflects the disproportionate number of conservatives among US climate change skeptics. Typically, identity dynamics dictate that skeptics who lean conservative express greater confidence in leaders with the same political stands and those whom they view as members of their in-group.[34] In the same vein, skeptics view mainstream and liberal media as a clear and threatening out-group that perpetuates the scientific narrative of human-caused climate change, a narrative that is particularly threatening to the skeptic identity.

In terms of other information sources that generate data related to climate change, skeptics who took our online survey expressed comparatively higher levels of trust in some select scientific organizations (e.g., 72.5 percent express trust in NASA and 64.9 percent for NOAA), while at the same time reporting lower levels of trust in others. Skeptics are particularly distrustful of scientific organizations affiliated with regulating climate change (only 50 percent express any trust in the EPA).[35] This trend has been noted elsewhere in the environmental sociology literature, where skeptics' distrust in science has been found to be more severe when it comes to "impact science" (science examining the effects of economic activities and technologies on the environment and health) over "production science" (science that leads to innovation and economic growth).[36]

Clearly skeptics have significantly low levels of trust in mainstream media compared to others from which they may also receive climate change information, including political leaders who comment on the issue or some limited scientific organizations that they trust. The climate change denials from trusted Republican political leaders such as former president Trump are especially concerning, given that most skeptics exist in echo chambers of like-minded individuals and are less likely to be exposed to news directly from scientific organizations; this political commentary by and large shapes their views about climate change.

Nuances of Trust/Distrust in Media

Previously, media scholars have pointed to two types of bias in the media that can lead to public distrust: "ideology bias," which refers to "a news outlet's desire to affect reader opinions in a particular direction"; and "spin," which "reflects the outlet's attempt to simply create a memorable story."[37] In other words, ideology bias is the traditional left or right partiality that comes from the preferences of editors and reporters, whereas spin is prejudice arising from media's attempts to, for example, simplify and make a story more appealing through discarding contextual information.

In their interviews, skeptics alluded to their ability to perceive both types of bias in media coverage of climate change. Interestingly, though, the way skeptics apply these concepts to media coverage of climate change is unique and different from how they are used by the nonskeptic public. Consider Blake, a white man with some college education who identifies as politically unaffiliated. On ideological bias, Blake states that mainstream media outlets predetermine what stories to cover and how to cover them, largely based on the outlet's political agendas and policy goals. He believes that this is true for all mainstream media, "except for *Fox News*." Blake continues:

> It [mainstream media] is certainly pushing climate change, except for *Fox News*. Like it is so far off on reporting unbiased scientific discovery relating to anthropogenic CO_2 driven climate change. There is no correlation in my mind. Like what the media is putting out and what the cutting-edge scientists are discovering, that is in opposition to what the media is putting out, there won't be any correlation. In general, what is being published on is pro-anthropogenic climate change, and so there is a bunch that the media can pick from. Are they picking at, you know, this is the best paper written on this topic and this is a recent discovery? No, their filter basically is, will this create fear, will this drive forward our agenda?

Given perceived biases and hidden agendas, Blake concludes that mainstream media is "a completely failed system" in terms of being able to "inform the public on what is going on so that they can have informed debates." To Blake the fact that one can so easily identify political affinities of mainstream media is one big reason to be suspicious of their coverage of controversial issues, including on climate change. However, Blake does not seem to apply this same critical lens to the Fox Network, which he recognizes as politically biased, yet alludes to trusting more for his climate-related information. "The fact that you can identify *Fox News*'s bias, and if you can identify an entire media station's political bias, it is like, Who told them to have that political bias? It came from somewhere." That Blake applies a generous interpretation to Fox's coverage of climate change while recognizing its clear political affiliation suggests that, driven by ideology and identity, skeptics such as Blake prefer news sources that enhance their values and worldviews.[38]

Citing various additional reasons, such as a profit motive, almost all skeptics we interviewed largely agree with Blake in believing that mainstream media is biased, driven by hidden agendas, and not trustworthy. These views among skeptics indicate a generalized distrust of the media and a significant alienation from mainstream news outlets, which can be especially problematic when trying to convey information on topics such as climate change that are politicized and where public views are already polarized.[39]

When skeptics refer to bias in climate change coverage they are primarily referring to a perceived ideological bias. Specifically, our research shows that skeptics believe climate change facts are distorted in the news to conceal accurate scientific data to mislead, scare, or control them; to suppress conservative viewpoints; or to impose liberal policy solutions on the wider populace. In truth, though, the "real bias" in media reporting of climate change is not an "ideological bias" but rather an "informational bias."[40] Accordingly, journalists' and media corporations' attempts to maintain the professional ethos of balance leads to a divergence from scientific consensus—resulting in coverage that emphasizes uncertainty about the issue.[41]

Besides this ideological bias, several skeptics also express concerns about spin as a reason for their distrust, though they used the term "sensationalism" more frequently. To Henry, a conservative white man from southern Idaho, profit motives lead the media toward sensationalism and a blurring of the truth: "So what produces money for the media? Sensationalism. And they could be accurate, but I bet you there are some people in the media that don't care if it is accurate. They want a buck . . . So you have to look at what they are doing and why . . . You have to look at the data and challenge everything."

Nick agrees and elaborates on why media sensationalism is problematic and insidious, concluding, like Blake, that the whole media ecosystem is "starting to collapse" due to nefarious practices and sensationalism that blur the line between "actual news" and "fake news":

> The media is just—there's a natural bias toward sensationalism because they want people to read their papers and, if they don't do that, what occurs? You end up with outlets like Buzzfeed creating even more sensational things, cutting into their audience. So, it's like, "We have to be sensational enough, but not so sensational," and so they're in this weird balancing trick trying to keep viewers—it's like, yeah, the media itself is starting to collapse especially, 'cause now instead of having a tabloid news section that people will pass in the store, now they have to compete with Buzzfeed or anyone creating a new Buzzfeed, to just push sensational things out in front of people whenever.

"Fake news" is a term that became popular during the 2016 US presidential election. The term primarily refers to intentionally fabricated and/or inaccurate new pieces, but there is no clear single definition of the term and it has been used in multifaceted ways to mean everything from "fabricated news circulated via social media to a polemic umbrella term meant to discredit 'legacy' news media."[42] Nick's and other skeptics' interviews reveal that "fake news" in the context of climate change means a number of different things; on the one hand, it implies media bias and sensationalism, which leads to inaccurate reporting or outright fabrications; on the other, it implies that all mainstream media peddle in lies and self-serving practices and are therefore untrustworthy.

That mainstream media sensationalizes content is a widespread perception among skeptics. This view was expressed to different degrees in almost every interview we conducted. To Nancy, media "exaggerate[s] it [climate change] or dramatize[s] it, is probably a better word." To Jane "some of them overblow it." Extremely low levels of trust in mainstream media, coupled with actual and perceived media biases, makes it particularly challenging for news outlets to cover stories about climate change *and* at the same time be perceived as objective, accurate, and credible. Climate news confronts skeptical views, not just those based on the science behind climate change, but also ones based on the perceived motivations of organizations that carry the news. Embedded largely in conservative echo chambers, skeptics, especially those with more salient intersecting identities (e.g., political identity) are more likely to reject coverage that is inconsistent with their views when the coverage is seen as politically biased or intentionally sensationalized to serve the political goals of the perceived out-group.[43]

ENGAGEMENT WITH INFORMATION / DISINFORMATION

Skeptics' distrust of media shapes how they engage with climate change information. In short, skeptics feel that, given the biases described previously, the public at the receiving end should be highly skeptical of any material presented to them and should regularly peruse "multiple sources" to expose themselves to "multiple viewpoints" before forming their opinions. Further, skeptics seek climate reporting that is "fact based," not "opinion based."

In the surveys we asked how respondents determine whether a news story on climate change is reliable. Answers range from "the channel airing the news" (approximately 30 percent of the sample), to "who is funding the research reported" (38 percent), to "the credentials of the scientists featured" (45 percent), but the dominant response is whether the news outlet "provides multiple points of view" (approximately 60 percent of the sample). Of all survey respondents, 64 percent also agree with the statement that "News agencies should allocate equal time to cover all of the different viewpoints about climate change."

These viewpoints are related to skeptics' overall impressions about media coverage of climate change. It is likely that those who assume ideological bias in the media prefer channels that are seen as neutral or balanced. It is also likely that skeptics' concerns about the credibility and trustworthiness of climate scientists drive their engagement with news stories about climate change. On the other hand, the preference for equal-time allocation for multiple viewpoints, though it seems reasonable, stems from a misperception of balance related to climate change news. As noted earlier, when overwhelming evidence-based consensus exists, allocating equal time to cover "both sides of the story" actually creates an informational imbalance by overreporting on scientific uncertainty.[44]

Almost all interview participants mention that they do indeed examine diverse sources and give attention to multiple perspectives on climate change before forming their own opinions. When we asked where she receives information on climate change, Jodie, a woman who identifies as Republican, states, "I go to—I like to look at almost every news source. Like CNN, *Fox News*, what is the other one? ABC. So just, kind of, for me, it's just like all over the place. And I listen to a lot of political podcasts. And, so, kind of what the politicians are saying about climate change. And then I look at some videos on YouTube about what scientists are saying. So I kind of just get it from all over the place." Jodie further exclaims, "And every time I read something, I make sure to see what the opposite side thinks about it." Jodie feels that this way of engaging with climate change information gives her a broader understanding.

Savanah, a white woman from north central Idaho, confirms this pattern by stating, "I use them all [media sources]. I'm the kind of person that I like to see all angles. If I see something on one news channel or on one newspaper, I have to read the other one. I have to read the flip side. You know? Read something from the liberal view, [then] I have to read it from the conservative's view. I don't just stick with one. I don't have one favorite over the other, either. I like to read it all. Then I form my opinion." Savanah's response not only affirms the importance of the "multiple source–multiple perspective" approach to media engagement but also hints that this might be one way to tackle perceived ideological biases of media.

For Jodie, Savannah, and other skeptics, their reasonable expectation of balance as a journalistic value seems to drive them toward a multiple source–multiple perspective approach to media engagement, which in fact exposes them to a higher amount of coverage that elevates their level of uncertainty and strengthens their skeptic identity. It is plausible to think that these well-intentioned media engagement patterns confront confirmation bias when new information does not align with preexisting viewpoints, which may drive individuals to seek out the "other side," which in turn provides them with a reasonable justification for rejecting the new information.

During his interview Sam explains both the challenge of confirmation bias and the need for balance: "I don't want to just read articles that confirm my bias. So, I really like reading Huffington Post and Salon and Vox because I want to get a different perspective as well. And make sure that, in my chest, kind of listening to a sounding board. Where just my ideas are being sent back to me, I'm never going to be a challenge to my way of thinking, so I think unfortunately that can happen."

Only one of our interview participants expressed any critique of this multiple source–multiple perspective approach. Contrary to his skeptic counterparts, Zed does not necessarily make an effort to expose himself to multiple views: "If they're talking about something that I don't believe in, I'm going to change the channel, so I guess I don't necessarily balance things out." In this way Sam and Zed both acknowledge confirmation bias, with one actively fighting against it and the other succumbing to it.

Trying to achieve balance using a multiple source–multiple perspective approach to media engagement receives overwhelming support from our survey and interview participants, which indicates that climate change skeptics are especially vulnerable to rhetorical strategies of the denial countermovement. For instance, this balance-seeking, from the perspective of the skeptic, may hinder the ability to distinguish comments from a qualified expert versus comments from someone lacking the credibility to discuss climate change, or to distinguish between climate science and pseudoscience. These are two denial techniques (among many) that John Cook has identified in his work on developing a typology of "denialist rhetorical strategies."[45] For example, a skeptic may, inadvertently, equalize the information coming from a reputable climate scientist with that coming from a skeptic blogger, which is especially problematic online, where there is a certain flattening of information. Online, a savvy blogger may appear equally or perhaps even more credible than a researcher whose life work is to examine and report facts on climate change.

Furthermore, the search for balance and multiple perspectives also increases the likelihood that skeptics are exposed to climate change disinformation (intentionally disseminated false information).[46] Disinformation originating from the denial countermovement is particularly hard to combat, as propagators manifestly target skeptical beliefs, identities, and ideologies. Moreover, genuine misconceptions are often hard to distinguish from intentional disinformation. Research suggests that those who propagate disinformation use common psychological biases (such as logical fallacies) to build denialist rhetorical strategies and techniques. Such techniques include, for example, attacking a person (e.g., scientist) or a group rather than engaging with their arguments, using anecdotes to refute scientific evidence, or demanding unrealistic levels of scientific certainty.[47]

When engaging with climate change information, skeptics also argue that they prefer facts over "opinions" or "emotions." Jennifer, a white woman in her thirties, seeks "more of the facts than the opinions of everybody else." Likewise, Savannah feels that facts and opinions should be clearly labeled as such:

> I do feel, very strongly, that it needs to be labeled "This isn't a personal opinion, this is certain. This is a global funder's opinion. This is not an opinion; this is a fact." I feel like that should be something that is labeled clearly, for everyone to see. Because if somebody knows it's a

personal opinion, and it's broadly broadcasted as a personal opinion and not just truth—I think that's something that needs to be clarified a lot.

It is worth considering whether what skeptics dismiss as "opinion" constitute statements based on inconvenient facts (e.g., scientific facts that do not align with their dominant views about climate change) or driven by confirmation bias, or a need to avoid information that may cause them emotional distress (the "ostrich effect," a topic addressed in chapter 6).

CONCLUSIONS AND IMPLICATIONS

Our data suggest that distrust of media, mainly of mainstream media, shapes how skeptics engage with climate change information as well as how they think information should be accessed and evaluated by the audience. Yet, despite their advocacy for a multiple source–multiple perspective approach, our findings suggest that most self-declared climate change skeptics exist within their own conservative echo chamber. They rely primarily on conservative media such as the Fox Network, talk radio, conservative blogs, aggregate news sites, and social media–filter bubbles for their information on climate change. As a result, they are disproportionately exposed to media coverage that is highly dismissive of climate change and distrustful of climate science and scientists, or both. Further complicating the issue, the primary information habit that skeptics advocate for, the multiple source–multiple perspective approach, is time-consuming, impractical, and potentially unproductive, given how difficult it is for a layperson to grasp all the complexities and uncertainties involved in climatology or how difficult it is to understand raw data related to climate dynamics when cross-checking primary sources. This approach also requires audience members to actively fight against psychological biases such as confirmation bias, which, by the admission of skeptics themselves, is no easy feat.

In fact, according to political scientists Charles Taber and Milton Lodge, confirmation bias is especially acute when it comes to political beliefs and especially among those with strong prior beliefs.[48] For a highly politicized topic such as climate change, this suggests that, regardless of open-mindedness to the information-processing that is advocated by skeptics themselves, people with strong skeptic identities have a harder time avoiding "motivated skepticism" (the tendency to seek out confirmatory evidence and counterargue contrary arguments).

It may also be the case that participants report a reliance on multiple, cross-ideological sources due to what researchers call "social desirability bias," or the tendency to respond to interview questions in ways that one believes is socially acceptable. Research on news browsing habits conducted by statisticians Seth Flaxman, Sharad Goel, and Justin Rao finds that "individuals generally read publications that are ideologically quite similar, and moreover, users that regularly read partisan articles are almost exclusively exposed to only one side of the political spectrum."[49] So, while our participants tell us they are reading information "from all sides," scholarship suggests few people actually engage with cross-ideology media.

While we know of no prior research that directly applies social identity theory to news consumption on climate change, this may be a fruitful avenue for future research. Scholarship does suggest that partisans (both Republicans and Democrats) are more likely to believe and share news that reflects on their in-group favorably and on an out-group unfavorably.[50] When it comes to skeptics, this in-group/out-group bias likely extends to media and information sources that disseminate the scientific consensus on climate change, a narrative that skeptics view as especially threatening to their group identity. As a result, skeptics may prefer news that aligns with their in-group attitudes, such as those emanating from "the denial machine," and express greater distrust regarding news that challenges their in-group attitudes.

Overall, our findings show that partisan media play a significant role in creating, strengthening, and sustaining opposition to climate change. While scholars elsewhere have reported that to reach people who are closed off to new information reporting on climate change should be carefully cultivated to its specific, targeted audience, our work with self-declared climate change skeptics suggests that individual-level persuasion techniques (such as issue framing) might be less successful at changing skeptics' minds, especially the minds of deniers.[51] Instead, elite cues arising from *within* the conservative movement may be more effective at gaining skeptics' support for climate change policy. In this vein, Republican political leaders, commentators, and others in conservative circles have an important role to play, including in communicating climate science and policy with the public. One encouraging sign is the slowly increasing number of Republican leaders in the United States who accept climate change or openly advocate for policy solutions to mitigate the worst effects.[52] Some young conservatives are also starting to become climate activists, disseminating countersignals via traditional and social media.[53] Indeed, given the identity dynamics discussed in the preceding chapters, climate change skeptics may be more receptive to messages from these trusted sources. (We elaborate on these sources, their implications, and other avenues for climate change communication in the concluding chapter.)

six

THE EMOTIONAL LIVES OF SKEPTICS

IN HIS 2020 CLI-FI NOVEL, *The Ministry for the Future*, science fiction writer Kim Stanley Robinson introduces readers to Frank May, an American aid worker stationed in India.[1] In the fictional future in which the story unfolds, a future "almost upon us," climate change has affected all. India is suffering from yet another scorching heat wave. The story begins with a description of a single morning in what has become Frank's everyday life:

> *Curious*, *alarmed*, feeling himself breathing hard, Frank walked down streets toward the lake. People were outside buildings, clustered in doorways. Some eyed him, most didn't, distracted by their own issues. Round-eyed with *distress* and *fear*, red-eyed from the heat and exhaust smoke, the dust. Metal surfaces in the sun burned to the touch, he could see the heat waves bouncing over them like air over a barbeque. His muscles were jellied, a wire of *dread* running down his spinal cord was the only thing keeping him upright. (emphasis added)

In our nonfictional world of today, accounts of climate change are starting to look and feel disconcertingly similar. David Wallace-Wells, a well-known American journalist who has written extensively about climate change, begins his own 2020 book, *The Uninhabitable Earth: Life After Warming*, with a similar avowal: "It is worse, much worse, than you think."

In both accounts, the writers use extensive emotion-based language to describe the devastation brought about by climate change and our individual, emotionally charged responses to it. And, with good reason. Survey data from the Yale Program on Climate Change Communication shows that two-thirds of Americans are indeed worried about climate change.[2]

Parallel to this increase in concern among the public and the increasing urgency in popular writing, there has been an increase in academic scholarship devoted to the intersection of emotion and climate change. This work is so extensive that scholars have constructed new language to describe the powerful feelings people experience towards the climate crisis: eco-anxiety, pre-traumatic stress syndrome, and ecophobia, to name a few.[3] Scholars increasingly argue that emotion should be considered as an integral part of a broader climate change communication strategy.[4]

While said scholarship is rich in theoretical and empirical investigations into how emotions drive public engagement with climate change, little attention has been paid to the role that emotion plays in climate change skepticism and denial. Do skeptics feel more or less of certain emotions when they think about climate change? If yes, what are those emotions? How do they differ from the affective responses of those who accept the veracity of climate change?

Our research suggests that, in general, skeptics experience less of an emotional charge around climate change than those who accept climate science. The skeptics we interviewed shared their neutrality, saying things like "I don't really care about it" (Sabrina); "When I think about humans causing global warming through CO_2, I feel no emotion" (Blake); "Well, I'm not a real emotional person. At all. Um, I'm level-headed, so . . ." (David); and "I don't get emotional about it I guess" (Mark). The lack of strong emotions about climate change among skeptics may be obvious, since it is natural to not feel anxiety or fear over an issue perceived to be nonexistent. However, when we asked these seemingly "unfeeling" skeptics to elaborate on their responses, we witnessed a rich emotional array. Skeptics express anger toward out-groups (e.g., scientists, liberals, environmentalists) and they worry about and dread things like animal species destruction, natural disasters, and future generations, including their children and grandchildren, "if" climate change is real.

Here we focus on developing a more nuanced understanding of skeptics' emotions toward climate change. To examine the relationship between climate skeptic identity and feelings about climate change, we begin with an exploration of social science scholarship on climate change emotions. We then examine three emotions—anger, worry, and dread—as experienced by skeptics, within a skeptic identity. We show how skeptics affirm their identity via anger and out-group derogation, while expressing worry and dread regarding specific "objects of care." We conclude by identifying potential implications of skeptics' emotional experiences for climate change communication.

CLIMATE CHANGE EMOTIONS

Recent social science analysis has paid close attention to the role of emotion in climate change opinion formation, policy support, mitigation action, and communication. Despite emotions' fuzziness and difficulty to measure empirically, they are essential for both practical and moral decision-making.

Neuroscientists and psychologists who study information processing employ the "affect heuristic" as a theoretical framework when discussing the role of emotions in decision-making. Paul Slovic and his colleagues have defined affect as "the specific quality of "goodness" or "badness" that is experienced as a feeling state (with or without consciousness) and which demarcates a positive or negative quality of a stimulus." In essence, the affect heuristic is the process people use to take mental shortcuts in their decision-making by relying on their affect and emotions. The heuristic practice lends itself to a quicker, easier, and more efficient way to navigate through a complex, uncertain, and sometimes dangerous world."[5]

Evidence from cognitive psychology suggests that affect and emotion are crucial for decision-making because both play an important role in motivating behavior. For example, when an individual faces an emotionally significant event, if a similar event stored in the memory gives rise to pleasant feelings, the event motivates action to reproduce the same feeling. But, if the event generates unpleasant feelings, it will motivate action that avoids the said feeling.[6]

Questions remain regarding when and how certain emotions motivate people to act on issues like climate change. Some researchers argue that fear tactics (e.g., framing climate change as the cause of negative health effects or as economically devastating) may provoke "denial, passivity, and fatalism."[7] Environmental social scientists, including Kari Norgaard, have cautioned that focusing on the dangers of climate change or a dystopian future may activate psychological defense mechanisms (e.g., denial) to mitigate the resulting emotional discomfort.[8] Others have argued that painting too rosy of a picture in the face of the prevailing climate crisis (for instance, a potential low-carbon future), may also be counterproductive as doing so ignores potential losses and leaves us "unable to work through our guilt and grief," further trapping us in denial.[9]

Even the impact of fear is contested. Writers like environmental journalist David Roberts have focused on leveraging the potential motivating effects of strong negative emotions such as fear, anger, and resolve and argue that embracing fear as a rhetorical strategy could motivate action.[10] Other scholarship suggests that other emotions such as guilt, regret, and self-accountability have proven to be variably effective at motivating action, alongside prosocial emotions such as hopefulness, pride, and gratefulness, although the application of these emotions to climate change have been inconclusive or contradictory.[11]

Casting a broader look at the extant literature on climate change emotion, in their 2017 *Nature Climate Change* commentary, Daniel Chapman and his colleagues argue that the "go positive" or "go negative" bifurcation in communication strategy is an oversimplification of what research on emotions reveals. According to these scholars, the either-or strategy overcomplicates the "very real communications challenge advocates face by demanding that each message have the right 'emotional recipe' to maximize effectiveness." Chapman's team advocates for considering emotion as "one element of a broader, authentic communication strategy" rather than a "simple lever" to be pulled in a specific direction in search of a specific outcome.[12]

We previously have argued that a potential "ostrich effect" may exist among skeptics.[13] The concept comes from research in finance and medicine, where the psychological discomfort of learning new information about potential future negative outcomes may lead to information aversion. Applying this to climate change skepticism, skeptics may be so driven to avoid learning about the dire consequences of climate change they will construct a safer alternative narrative, like the belief that climate change is a hoax. In short, for skeptics, climate change skepticism may act as a psychological coping strategy. Given this tendency, a communication strategy that centers on providing additional information is bound to fail.

Anxiety, Worry, and Dread

Much recent scholarship about climate change use different and newer terminologies, such as "ecophobia" and "ecoanxiety," to refer to varying degrees of concern regarding environmental problems.[14] In her book *A Field Guide to Climate Anxiety* environmental studies scholar Sarah Ray defines ecoanxiety as "a feeling of dread about the future combined with a feeling of powerlessness to do anything to shape that future."[15]

In work with our collaborator and environmental humanities scholar Jennifer Ladino published in the journal *Emotion, Space and Society*, we built on extant work on fear to suggest that climate change anxiety, worry, and dread are distinct emotions. Specifically, because climate change taps into existential concerns about "suffering, nihilism, and mortality," fears about climate change can manifest as dread, which we describe as a "sinking feeling, a paralyzing weight in the chest or stomach that is more intense than anxiety."[16] In contrast, worry can manifest as a drawn-out emotional state and is often tied to specific objects. While worry is often used synonymously with anxiety, we describe anxiety as a background feeling that is not easily identifiable with an object.

In that piece we further demonstrate that skeptics do feel worry; however, their worries are tied to specific "objects of care." Objects of care include the people, places, and species that we worry are affected by climate change (as compared to worry about climate change itself). Even skeptics express concern when it comes to objects of care that are meaningful to them, while rejecting or remaining uncertain about the larger phenomenon at play. We proposed a new definition of worry: "an affective state that directs anxious feelings toward particular objects of care without succumbing to the pitfalls of fear."[17]

Elsewhere, the broader environmental sociology scholarship suggests that liberals, nonwhites, and women express higher levels of worry about environmental issues compared to their counterparts. Notably, conservative white men have significantly lower levels of worry regarding environmental issues compared to all other American adults.[18] Worry is also a stronger predictor of climate change policy support when compared to disgust, hopefulness, anger, sadness, interest, and guilt. As a result, worry is better suited to motivating climate action.[19]

Anger

While climate change can provoke strong negative emotions such as fear, sadness, and anger in almost anyone, the origins of anger may vary based on whether one accepts the veracity of anthropogenic climate change or is a skeptic.[20] Skeptics may feel anger because they believe climate change simply does not exist and thus feel ire toward those who perpetuate the narrative that it does. Those who accept climate change, by contrast, may feel anger due to perceived climate inaction. To date, extant scholarship on anger has largely focused on the general public, with little attention paid to those who are skeptical of climate change.

Some studies on anger explicitly limit their samples to only those who accept climate science in order to clearly differentiate the causes of anger or to focus on the adaptive nature of anger.[21] For instance, in their work on "eco-emotions" (emotions evoked by ecological crisis), psychologist Samantha Stanley and her colleagues have uncovered that, for climate change accepters, eco-anxiety and eco-depression do not motivate action but "eco-anger" does. Accordingly, eco-anger is associated with higher levels of pro-climate behaviors, making anger an important emotional driver of engagement with the climate crisis."[22]

In one framing study, communication scholar Teresa Myers and her colleagues examined the effects of environmental, public health, and national security framing of climate change responses. They found that national security framing generates the most anger among those who are dismissive or doubtful of climate change. While Myers and her colleagues could not ascertain the causes of this anger, they speculate that the anger may have been directed at the researchers for presenting news stories that connect climate change and national security, two issues that skeptics do not view as connected. Indeed, skeptics may have felt that the framing article coopts values they care about.[23]

One of the few studies that does shed some light on potential causes of anger among climate change skeptics comes from a field that might seem unlikely, artificial intelligence. In that study Chieling Yueh developed an algorithm to examine how two opposing Reddit communities, r/climate and r/climateskeptics, discuss climate change. They find that though the r/climate community uses more sadness-related terms, between the two communities there is no significant difference in the use of anger terms. This potentially indicates that anger is used by both skeptics and accepters, but for different reasons. Interestingly this work also finds that the r/climate community uses more first-person pronouns (I, me, mine, we, us, our), which potentially reflects group cohesion regarding collective action. In contrast, conversations including second-, third-, and impersonal pronouns (you, your, he, she, they, them, it, anybody) are more commonly found among r/climateskeptics community, suggesting that these claims primarily reflect views toward the out-group. As such, Yueh concludes that the "r/climateskeptics community is more conscious about their opposing group (supporters of the consensus of climate change)" and gives more attention to the out-group.[24]

Identity theorists shed additional light on the topic of anger. Research has shown that when an individual's prominent identity is threatened, they may experience a negative emotional response such as distress, anger, or hostility.[25] The strength of the emotional response is tied to the "salience hierarchy" of the individual's identities, with more important identities producing stronger emotional reactions.[26] More intense anger is experienced in situations where *group-based* identities are not verified, whereas less intense anger is experienced with nonverification of *role-based* identities. For example, one may experience more anger when experiencing a challenge to their family identity (a more intimate group-based identity) than when faced with a challenge to their worker identity (a less intimate role-based identity).[27] Interestingly, when intense negative emotions such as anger are experienced, if the nonverified identity is salient to the individual's overall self-concept, a coping strategy may be to engage in behavior that affirms and reasserts the original identity.[28] In the context of climate change skeptics, these works suggest that when the skeptic identity is challenged, the more salient the identity (for example, denier versus doubter), the more likely the individual will experience anger and engage in behavior that solidifies their identity as a skeptic.

ANGER AMONG SKEPTICS

Let us consider Blake, the politically unaffiliated white man and former Libertarian from northern Idaho. Blake believes that all political parties are "manufactured" and should be "more or less dissolved or at least challenged." Blake argues that those who accept climate change are "egocentric" and "more than a little bit influenced by the agenda that they wouldn't realize that they are influenced by." Blake claims to feel "no emotion" when thinking about anthropogenic climate change. Yet he also conveys more than a little anger when describing how climate change "propaganda" is being used to cause "emotional distress" among the public.

> When I think about humans causing global warming through CO_2 I feel no emotion. And I think that the way that the agenda is being propaganda, it causes emotional distress in the same way that other things that are being propaganda causes emotional distress. When I think about the individuals that I know who are extremely passionate about this and would consider themselves climate advocates and that sort of thing, I feel much the same way about friends who are being tricked into buying a house [during] a housing [market] bubble or tricked into buying a really expensive SUV. And it is like "Why did you guys get sold on that?" or, you know, whatever. Friends can be deceived on anything.

Similarly, Mark, a Libertarian from northern Idaho, is "appalled" by how climate change rhetoric is being used to mislead the public. "I remember one, oh years ago, I guess, my in-laws watched the Al Gore movie, and they came out to visit and they were all wound up and thought the ocean was gonna come into their lake within a few years. They are pretty intelligent people too. I was kinda appalled." According to both Blake and Mark, the deception is so strong that it causes significant emotional stress to believers. This angers them, especially when those who are deceived happens to be their close friends or family.

While those who accept climate change may experience anger as a result of perceived inaction, for skeptics, anger emerges because they believe climate change does not exist and is being used to deceive or cause fear among the public. Skeptics feel anger toward those who say they believe in climate change but do not make the necessary lifestyle changes that they themselves advocate for and toward identified out-groups, including climate scientists and environmental activists.

For the most part, though, explicit discussions of emotions appeared rarely in our face-to-face interviews. Many of our interviewees responded curtly to our question about climate change emotions, often with a version of "I don't care" or "I don't get emotional about it." This is not surprising when considering the categorical identities that intersect with a climate skeptic identity. Skeptics in the United States tend to be mainly conservative-leaning men. Early sociological scholarship suggests that men are less emotionally expressive in general, due to the effects of gender socialization and the politics of masculinity.[29] Where men do express emotion, it is often evoked in the form of anger rather than, for instance, sadness or worry.[30]

Much of our more insightful data about skeptics' anger come from the survey. In the survey we asked, "While thinking about the concept of climate change, to what extent do you feel the following emotions?" Response categories included anger, disgust, calm, worry, dread, sadness, and grief, all measured on a Likert scale from "not at all" (1) to "an extreme amount" (7). Subsequently we provided respondents with an open-ended question to "explain why you feel the above emotions when thinking about climate change."

These open-ended responses uncovered numerous expressions of anger. For most skeptics in our sample, anger emerges from the belief that anthropogenic climate change simply does not exist. They instead believe that people with various nefarious motives use this largely fictional concept in self-serving ways: to mislead and control the public, to cause the public emotional distress, or to gain various financial benefits. Consider the comments that follow, along with demographic information that we could gather as indicative of each respondent's positionality.

> My anger is directed at the actual concept . . . it was Global Warming with panic association, now it is Climate Change so as to give it a more calming association but still associate it with devastation in the near future. It is a joke . . . climate is moving, migrating if you will. What we have in our generation will be somewhere else in future generations . . . and there is nothing we can do about it other than

be reasonable stewards of our current environment! (white man in his seventies)

I get angry at the manipulation of people's fears. (white man in his seventies)

Since nothing "climate change" fanatics have predicted has ever come true, I feel disgust, anger, and sadness over the concerted effort by these people to push this failed concept onto our population and particularly to propagandize our students in public schools with this myth. (white man in his seventies)

I am not worried about man made climate change at all. It makes me feel annoyed that so many people have made huge amounts of money on it. I feel annoyed and angry that children are being indoctrinated wrongly about it. God is in charge of the world. He will change the climate as He sees fit. Man has no power over it, only to the extent that we should take care of it by keeping it clean. (white man in his forties)

Because I've seen the hoaxes (overpopulation, ozone hole, etc.) over the past 55 yrs and find it disgusting the way they use/deceive people to further their own agendas. (white woman in her seventies)

I DON'T [worry]. YOUR [*sic*] AN IDIOT. CLIMATE CHANGE IS BS FAKE NEWS. (white man in his eighties)

So many are in this for the money and so many are just stupid . . . I believe that this has all happened before thousands of years ago and there is nothing we can do to change it and that is what most say. (white man in his seventies)

I feel angry that the MSM [mainstream media] use it as a tool for the left to push their Globalist agenda and try to scare people into following their sick leftist agenda. (white/Native American man in his forties)

These expressions of anger appear to be tied to conspiracy ideation or religious ideation.[31] For skeptics who believe that climate change is a hoax, anger is directed at those whom they perceive as perpetuating this hoax and at all members of deceptive out-groups, such as the mainstream media identified by one of the participants.[32]

Social identity theory tells us that individual views among skeptics are shaped more by in-group allegiance (and associated out-group derogation) rather than mere fact. Our data shows that when contemplating the topic of climate change, in-group allegiance works to provoke anger toward those constituting the out-group. In addition to mainstream media, targets of skeptics' anger include scientists, journalists, politicians, environmental activists, and celebrities. It is plausible that messages and policy proposals emanating from these groups feels threatening to and evokes anger in a climate skeptic identity. In particular, skeptics in our surveys directed their anger toward scientists whom they feel harbor hidden motives, including the desire for power, money, and other incentives.

> Scientists anger me and disgust me. (white man in his seventies)
>
> It angers me because the driving force of some "climate scientists" is purely political or to gain power and notoriety. (Arab American/Middle Eastern woman in her fifties)
>
> I don't worry about it but am disgusted with the corruption and greed of scientists and hypocrisy of advocates and politicians who support it. (white woman in her fifties)

Skeptics also experience anger toward climate change accepters based on what skeptics perceive as accepters' hypocrisy. They are particularly angered by the fact that accepters' actions do not align with behavior that they themselves advocate for as essential for curbing the worst effects of climate change. Overall, skeptics are highly critical of the behaviors of accepters, the out-group. Again, consider the survey responses below.

> I'm not worried about climate change. I do get disgusted with people who preach it but don't live their lives the way they tell the rest of us to live ours to so call[ed] save the planet. (white woman in her sixties)

> I think if everyone does their part it will help. But everyone that is crying and screaming don't do their part. Even climate protests [*sic*] they leave trash and drive cars to get there. What's the point. (white man in his seventies)

> So many idiots that are doing the talking but still are living in a manner that totally is against what they say or are just plain clueless . . . (white man in his seventies)

Anger among skeptics is so strong that some participants used this section of the survey to critique us, the researchers. They made such comments as "Bitch," "Fuck you," or, as mentioned, called us "idiots."

From early quantitative work on emotions, we know anger leads to polarization of climate change views between liberals and conservatives. Specifically, anger decreases policy support among conservatives while increasing support among liberals.[33] Early work has also shown that men are more likely than women to express climate change emotion via anger.[34] We examined the effects of gender and three ideologies—religious ideation, conspiracy ideation, and political ideology—on anger among survey respondents. The only factor that emerged as statistically significant was political ideology, with a small positive effect indicating that, among self-declared skeptics, those who lean conservative report feeling more anger when thinking about climate change. Gender, religion, and conspiracy adherence are not associated with anger in this context.

When we compare anger levels of those who deny human causes of climate change (deniers) to those who remain uncertain about the human contribution to it (doubters), the difference in anger levels is statistically significant, with slightly higher anger levels expressed by deniers than doubters (see chapter 7). This suggests that the more salient the skeptic identity is to one's self-conception, the stronger the negative emotional response expressed in the form of anger. In other words, deniers feel more anger than doubters and, for both groups, anger seems to be directed at clearly identified out-groups.

WORRY AND DREAD AMONG SKEPTICS

What may seem surprising, or even counterintuitive at first, is that skeptics like Savannah do evoke worry and dread to varying degrees when thinking about climate change. While identifying as a skeptic, Savannah expresses concern about the state of "our world." Her views indicate both sadness and anticipatory dread about the state of the world, *if* conditions were to get worse. When we asked her what emotions she feels about climate change, Savannah replies,

> Sadness. Watchfulness. I like to—I'm very watchful right now. I feel like there's a lot going on in our world. Our world is kind of starting to hit that turmulous [*sic*] stage, and What are people going to do? You know? Um, I think just waiting. Patience. I've got to be patient. There's going to be an opportunity to jump in and try to fix things and it's going to have to be a group thing, obviously. It's going to have to be a movement to really make a big impact on the whole world. But I feel like there will be an opportunity if things get bad enough. But, mostly sadness. It's really just sad to see the world get thrown into one extreme over the other when you kind of know in your heart that, if people would just find a balance between the two, they wouldn't be so extreme.

Other open-ended comments from survey respondents suggest that feelings of worry and dread related to climate change are not uncommon, even though skeptics may feel these emotions to a lesser degree than those who accept climate science. Consider the below statements from our survey respondents that evoke worry:

> I worry about the animals and their habitats. I'm upset that our environment is being changed due to fossil fuels and humans. (white woman in her thirties)

> I am worried for the earth and I want to do something about it but feel like I cannot. (white woman in her twenties)

It makes me worry about the way the world will be if it'll have trees, polluted Ocean, or whatever whenever my son is old enough to be able to enjoy what's left of it. (white woman in her fifties)

Its distressing and sad that we can't figure out how to live more harmoniously with nature. (white woman in her fifties)

I just hate to see nature as we know it go! (white man in his seventies)

And these statements that evoke dread:

I'm absolutely terrified that we have already pushed the earth past the point of return and will now be just trying to do damage control to avoid extinction. (man in his forties who identifies with "other" racial category)

It's scary not knowing if I can survive excess heat or cold. (white woman in her fifties)

There is so little time and so much to do. We are a species with amnesia because every generation has to learn it all again. I fear the human race is doomed. (white man in his forties)

Because it could bring death . . . because it could make or break this world as a whole or all kinds of bad things. (white man in his thirties)

Here we should make an important distinction between worry and dread. Dread is anticipatory, longer lasting, and generates a heavier feeling, as we demonstrated in our article published in *Emotion, Space and Society*.[35] When assessing the interview data in combination with open-ended responses from the survey, we find that dread among skeptics is mainly associated with feelings of existential threat, such as humanity's ultimate survival on the planet. Dread constitutes the idea that "the human race is doomed," as one survey respondents stated.

Worry is "more other-directed and is produced in response to consideration of habitat and nonhuman animal species extinction."[36] From the examples, worry refers to specific "objects of care" such as animals, habitats, the earth, and nature. Dread is "the point of no return." To further illuminate this, consider our interview participant Zed, a white man from northern Idaho who *worries* about carbon footprints and deforestation while expressing uncertainty about its long-term effects:

> Well, we're obviously adding a carbon footprint by buying more vehicles. Some say that the new vehicles that we are putting out [do] lower emissions, but we've almost doubled or tripled the amount of vehicles that are out there, so even if it's half, we're still putting the amount out there. We're cutting down forests. But we have a lot of reforestation programs, especially in Europe, Asia, and the United States, and now Africa, and the jungle is getting cut down pretty fast and pretty hard and it's not recovering. So where does the balance lie? I don't know, but we are making an impact on it. How much and what kind of impact I don't know.

Or Allen, another white man from northern Idaho who *worries* about pollution, particularly garbage in the ocean:

> Well, because even if climate change isn't real, there's still all the pollution and all the garbage in the ocean. It still affects people even if it's not changing anything in the ozone layer or anything like that. It is still affecting people in China, in those really populated areas where they have to walk around with those masks on and things like that.

Compare the above expressions of worry to those of one of our survey respondents earlier, the man in his forties who expressed his emotions in the form of *dread*: "I'm absolutely terrified that we have already pushed the earth past the point of return and will now be just trying to do damage control to avoid extinction." In short, these findings indicate that dread emerges from a heavier, deeper, and more unsettling feeling about anticipated futures, whereas worry emerges from concern about specific, more immediate events and objects such as animals, plants, habitat destruction, and species extinction.

In fact, our survey responses regarding the extent to which skeptics feel worry and dread about climate change reveal that 15.7 percent of skeptics express "quite a bit," "very much," or "an extreme amount" of worry, with twenty-six participants (2.6 percent) selecting the extreme worry category. Further, 11.5 percent express "quite a bit," "very much," or "an extreme amount" of dread, with twenty-three participants (2.3 percent) selecting the extreme dread option.

Given that our survey sample consists of skeptics, it may seem surprising to see any level of worry or dread regarding climate change, since we know that worry and dread levels are lower for those who deny human causes of climate change compared to those who doubt human contributions.[37] To elaborate, in our survey, average worry level for deniers was 2.29 in comparison to doubters, which was 3.11. This is a statistically significant difference. Further, average dread level for deniers was 2.09 in comparison to doubters, which stood at 2.70, again, a statistically significant difference. In other words, skeptics who doubt anthropogenic climate change experience more worry and dread compared to skeptics who outright deny its existence. (This is in contrast to anger, which deniers experience more than doubters, as already discussed).

When we further explore whether sociodemographic factors help explain these emotions, we find that, among skeptics, women experience higher levels of worry and dread compared to men.[38] This is not surprising, given the vast extant literature on factors that shape our emotional experiences. We exist in a cultural context where gender socialization promotes women's expression of emotions via sadness and men's expression of emotions via anger.[39] These gendered patterns are potentially amplified in our surveys with skeptics, the majority of whom lean politically conservative. Conservatives tend to hold more rigid beliefs about gender roles than do progressives.[40] These intersecting dynamics may explain the difference between skeptical men's and women's expression of worry and dread related to climate change.

Given the underlying ideological beliefs that shape our identities and experiences, we also investigated whether conspiracy adherence and religious beliefs shape skeptics' emotional experiences. Previously, in an article published in the journal *Sociology Compass*, we argued that conspiracy ideation potentially acts as a "psychological coping strategy," where conspiracies mitigate certain negative or uncomfortable feelings by providing skeptics with simple, clear explanations for large, complex problems.[41] Conspiratorial explanations often attribute the causes of one's discomfort to a single actor or event, hence reducing overall concern. As a result, skeptics who believe in the existence of a climate change hoax may feel less worry or dread about the issue. After all, why worry about an issue that probably doesn't exist? Similarly, religious ideation is known to reduce negative emotions such as worry and dread by increasing the senses of well-being and belonging for people who participate in religious communities and activities or by providing theological explanations for events that occur on Earth.[42]

When looking closely at open-ended responses from the survey, evidence of the moderating effects of conspiracy ideation and religious ideation emerge. Consider the following examples, which came from survey respondents who selected "not at all" regarding the extent to which they feel worry and dread about climate change. Some of these responses indicate a clear connection between conspiracy ideation and low levels of worry and dread:

> It's not a bother to me and is probably mostly fabricated. (white man in his sixties)
>
> It is a complete political hoax. (white man in his eighties)
>
> Climate change is the biggest hoax that was ever perpetrated against the people of the world. (white man in his sixties)
>
> The rediculus [*sic*] notion that all climate change is caused by humans. (white woman in her sixties)

Other responses indicate a clear connection between religious ideation and low levels of worry and dread:

> I think God is in control so I don't think any negative feelings help at all. (white woman in her seventies)
>
> I have no emotions concerning this subject. I know that Jehovah will fix these problems soon. (white man in his fifties)
>
> I do not believe in climate change. God rules over all. (Black/African American man in his fifties)
>
> Climate change as a whole is not happening now, but will in the future. I take the Biblical view that because men reject God and his son The Lord Jesus Christ, God will scorch men with heat during

the Tribulation when the Antichrist is here as foretold in the book of Revelation 16:5–9. (white man in his seventies)

Further statistical analysis of factors associated with skeptics' feelings of worry and dread reveals that skeptics who have experienced "negative environmental events" (in particular, direct and indirect experiences with water and air pollution) are more likely to express worry and dread regarding climate change.[43] There is a dearth of extant empirical work on the association between environmental experiences and emotions. In one thorough review of personal and social factors connected with environmental concern and behavior, Robert Gifford and Andreas Nilsson points to a number of factors that may be pertinent to understanding skeptics' emotions. Evidence suggests that children who spend more time outdoors, or young people who read environmental books or watch nature shows, develop greater concern about the environment as adults. Furthermore, proximity to "problem sites" (e.g., landfills and waste disposal sites) also leads to higher concern about environmental problems.[44] It appears that similar effects exist among skeptics who have experienced certain negative environmental events. While they reject climate change and its human causes, skeptics seem to worry about environmental impacts, especially when they are experienced directly. Taken together, our results suggest that skeptics' dread and worry is shaped by a number of factors including identity (e.g., gender, political orientation), ideology (e.g., religious ideation, conspiracy ideation), and direct experience (e.g., exposure to negative environmental events).

To understand why worry and dread matters, we need to understand the effects these emotions have on skeptics' concern for the environment and support for pro-environmental policies. We uncovered that, among skeptics, those who report a higher degree of worry and dread regarding climate change also express a higher degree of environmental concern and support for pro-environmental policies, such as investment in solar and wind energy and federal regulation of air and water pollution.[45] In short, skeptics who express more worry and dread about climate change are also those more likely to care about environmental issues and support policies that protect and benefit the environment.

Earlier we demonstrated that some skeptics harbor certain pro-environmental views, such as concern for pollution, habitat destruction, and species extinction (which we refer to as "objects of care"; see Wang et al. 2018). We extend this finding by showing that skeptics who harbor pro-environmental views also express higher levels of worry and dread about climate change. However, despite overwhelming scientific evidence showing a connection between climate change and environmental issues, skeptics do not necessarily connect their concern about environmental issues to concern about climate change.

This observation extends research by psychologist Susan Wang and her team. They suggest that having "objects of care" reduces the "psychological distance" between oneself and the world's climate (i.e., that climate change is occurring in temporally, socially, and geographically distant settings), making the issue "more personally relevant, evoking stronger emotions, and prompting action".[46] Our research shows that for skeptics, even though they deny or doubt human causes of climate change, their concern about "objects of care" still elicit worry and dread. In this case, the psychological distance between skeptics and climate change need not necessarily be bridged for them to care about certain environmental problems and to want to address them.[47]

BEYOND ANGER, WORRY, AND DREAD

Beyond anger, worry, and dread, the open-ended responses from our surveys and interviews provide tentative evidence for a number of other emotions that may also be at play in skeptics' emotional experiences. For example, Jodie, the Latina woman who leans Republican, stated that when she thinks about climate change, she only feels "curiosity":

> Mostly I think, like, curiosity. I don't feel angry. Or sad. I'm not angry like "We are destroying our world, we need to do something right now." Or I'm not angry like "People are deceiving people." It's more, I think, a neutral thing. I don't feel that urgency that some people do. Or that anger. I just feel like a curiosity. I want to know what both

sides are thinking. I want to see "Is there anything I can do to help?" I want to see, is there anything we can do to help? And so it's just, kind of just like, a neutral feeling, curiosity.

Jodie's description, while acknowledging the anger of both believers and skeptics, provides an interesting juxtaposition between worry and anger, landing at a neutral position: curiosity. It is plausible that others like Jodie who are uncertain about anthropogenic climate change, including those who have not paid much attention to the issue, express their feelings using emotion terms with a neutral valence, such as curiosity or interest, rather than anger, anxiety, worry, or dread. This topic warrants further investigation.

We also observe resignation among some survey respondents, who stated "There's nothing I can personally do" or "Why worry about something you cannot change?" Their lack of worry seems to be partly a result of a perceived lack of agency. Several survey respondents also express feeling "calm" in the face of a changing climate, for various reasons. Some believe "it's not true." Others are "not interested in the subject." Some believe "it is a natural aspect of the earth and unavoidable." Still others do not worry about long-term impacts and simply state, "I don't worry about what might hurt me later. I'm more concerned about what could hurt me now." Others "haven't been affected" and therefore are "not concerned." These, coupled with a slew of emotions that other emotion researchers have examined, such as disgust, hope, sadness, fear, and guilt, provide avenues of potential interest for future research on skeptics' emotional lives.

CONCLUSIONS AND IMPLICATIONS

In this chapter we extend prior research on climate change emotion by considering the emotional experiences of self-declared skeptics. Specifically, we highlight three distinct emotions—anger, worry, and dread—and consider the extent to which skeptics feel these emotions, what specific entities, events, or objects might elicit these emotions, and why emotions matter.

It is clear that emotions play a nuanced role in driving opinions and responses to climate change, including those of skeptics. A better understanding of skeptics' emotions related to climate change could be used to design messages that target not only skeptics' informational needs but also their emotional needs. Our work has several implications for climate change communication that appeals to skeptic emotions.

Considering worry and dread, our findings suggest that even those who are skeptical of anthropogenic climate change still exhibit worry about specific objects of care, like species extinction and habitat loss. Framing messages to focus on these objects of care may increase skeptics' engagement with certain environmental issues. By extension, policies that address objects of care may garner wider support, even among climate change deniers or doubters. While these communication strategies may not change skeptics' perspectives about climate change, they may increase support for pro-environmental policies that in the long run will help mitigate climate change.

Regarding anger, we see two main limitations with prior research. First, much of this research has focused on pro-climate segments of society (e.g., environmental activists) and the anger arising from their concerns about a changing climate (or inaction to curb its impacts). Very little research exists on climate change anger among skeptics. Second, survey work that measures anger through quantitative scales alone encounter limitations when interpreting results; it is hard to delineate the origins of anger or at whom/what the anger is targeted. Through our work we begin to address some of these limitations by focusing on a neglected segment of the American public when it comes to anger scholarship—self-declared skeptics—and, by combining quantitative and qualitative data, to develop a more nuanced understanding of anger in the context of skepticism.

These early stages of work suggest that skeptics with strong identities are more likely to evoke anger in the face of climate information stemming from out-groups. Earlier in the chapter we posited that when an out-group challenges an identity that is salient to a skeptic's self-concept, they may experience anger (or other intense negative emotions) that in turn leads to coping strategies that affirm or reassert their identity. This may mean that when scientists, journalists, and others present pro-climate information, skeptics who experience anger as a result may turn inward, further entrenching themselves in echo chambers of like-minded individuals with whom they then engage in out-group derogation. These mechanisms of anger and identity verification suggest that individuals with strong skeptic identities are unlikely to be receptive to climate change messaging coming from the "other side." However, as our work on worry and dread indicates they may be receptive to discussing matters regarding the well-being of objects of care.

It is worth noting that among our interview participants, gender and emotions align. Men primarily express anger. Women express worry and dread. Survey data affirms this, to some extent. Indeed, skeptical women express more worry than skeptical men (3.01 versus 2.52 on our 1–7 Likert scale). Skeptical women also express more dread than skeptical men (2.66 versus 2.23). Both differences are statistically significant. However, we do not find a significant difference for anger between skeptical men and women (2.48 versus 2.70).

This does not mean that there are no gender differences when it comes to anger. Both the origin of anger and the construction of the survey question ("While thinking about the concept of climate change, to what extent do you feel the following emotions?") may be relevant. It is possible that some skeptics' thoughts about specific out-groups (e.g., scientists, environmentalists, democrats) led them to choose a higher value as their response, while others may feel anger for our even suggesting that this is an issue worthy of study. The open-ended responses in the survey provide some hints that both may be the case. We agree with other emotion scholars: it is particularly hard to delineate origins of emotion and these topics require more nuance methodological applications and analysis in future work.

We also note that in our interviews and in open-ended survey responses, the subjects of climate change and issues around environmentalism are deeply conflated. Indeed, it is difficult to separate the two, for researchers and for skeptics alike. For example, we know that Monarch butterflies and bees are dying as a result of climate change. Is concern about their loss related to opinions about climate change or is it related to concern for the environment? The same can be said about coral bleaching or melting ice caps. Many factors contribute to climate change (e.g., overpopulation, pollution, deforestation) or are outcomes of climate change (e.g., drought, fire). To skeptics these might be separated from climate change in their evocation of emotions.

It is crucial that climate change communications aimed at skeptics acknowledge the role of emotion and integrate emotion as part of a comprehensive communication strategy. The evidence is clear: among skeptics some strong negative emotions such as anger correlate with low acceptance of climate change and low support for climate policy. Communication strategies should target specific audiences and tailor messaging to their values and concerns. For example, if the target audience is climate change skeptics who may be experiencing out-group hostility, anger, and distrust, a scientific lecture on health impacts may be less effective than a personalized story about a regional issue of value, such as the threat to the redwood forests of California.[48] Wendy Ring has recommended selecting "images of people who mirror the age and ethnic mix of your audience" when presenting information about climate change.[49] This may be a useful strategy when communicating with skeptics, since they are more likely to refute both the messengers and the messages from the "other side." Given the claims of our interviewed and surveyed skeptics, helping them identify more closely with those who are harmed by climate change, such as children or specific members of future generations, may also be an effective strategy. In other words, make the issue personal. However, since many skeptics question the reality of climate change, it may be futile to discuss its long-term implications. Rather, communication with skeptics should focus on the here and now, emphasizing our collective responsibility and ability to protect objects of care that they value.

seven

TOWARD A CONTINUUM OF SKEPTICISM

NICK, THE IDAHOAN WHO FEELS ALIENATED from his town, family, and politics in general, has strong feelings about (what he perceives as) the inaccuracy of climate science. In his interview he remarked that not all skeptics feel as he does; he offered important insights into the range of skeptic perspectives. He explains: "You have people who just outright deny what they are being told, you have skeptics who listen to what they are told and they are not certain about it, and then you have people on my category who looked at it, understand it and they're outright skeptical. So, I am at the furthest extreme that you can get." He continues: "I don't deny and I'm not a skeptic. I full out dissent on this." Later in the interview Nick doubled down on his identity as someone who dissents. "I don't deny. And I don't have any doubt anymore. At one point I did. I just outright dissent. It gives a little more fidelity to it. It (dissent) can ascertain someone who has been told things by reading it versus someone who has actually looked into it." Nick establishes layers of skepticism by distinguishing those who have done their own research about climate change *and* determined that it is not real from those who either do not have a clear sense of climate science and remain uncertain about the phenomena or those who blindly reject climate science.

Nick's explanation raises questions about the terminology used by scholars, politicians, and reporters when discussing climate change. Specifically, in academic research on climate change skepticism we see scholars using competing language to define people who fail to accept climate science.[1] McCright and Dunlap refer to people as deniers; Rahmstorf uses the term skeptic.[2] Dunlap later distinguishes between deniers and skeptics based on people's willingness to engage with information and potentially change their minds.[3] In our article "Considering Attitudinal Uncertainty in the Climate Change Skepticism Continuum" we offer the following terminology for scholars to use when discussing climate change skepticism.

SKEPTIC: an umbrella term including anyone who fails to accept climate science, to any degree, for any reason;

DENIER: skeptics who refuse to entertain or accept new information on climate change, who have closed their minds to further consideration of the issue;

DOUBTER: those who do not accept the science around climate change because they are either uncertain or refuse to take a position on this issue.[4]

Nick's statements also lead one to question how researchers approach climate change skepticism. He argues that skepticism is not uniform and that various degrees of skepticism need to be considered. Our previous work with skeptics supports Nick's viewpoint and suggests that approaching skepticism as a continuum is the most accurate way to capture the nuances of perspectives related to climate change skepticism.[5]

We are not the first to propose this approach. Dunlap, in his article "Climate Change Skepticism and Denial: An Introduction," suggests that social scientists should envision climate change skepticism as a continuum, from skepticism (those who do not believe climate change is a serious problem but remain open to new information) to denial (those who reject the science around climate change and are not open to changing their minds).[6] We extend Dunlap's continuum to include those who are unsure about climate change or refuse to take a position on it. The continuum then expands over a series of positions to end at those who, like Nick, actively and obstinately deny that the earth's climate is changing.

We have explored how climate change skepticism operates as a social identity and the relationship it has to an individual's perceptions of climate science and climate scientists, the way belief in conspiracies contributes to climate change skepticism, how religious ideas influence perspectives on climate change, skeptics' environmental beliefs and behaviors, the kinds of media and information sources skeptics rely on, and how skeptics emotionally respond to climate change. We now consider how the degree to which one is skeptical might influence, or be influenced by, all of these factors. We begin by presenting a series of points on a skeptic continuum to use for analysis, then create profiles of skeptics who fall along these points. We then use these profiles as case studies to analyze how the strength of skeptical beliefs relates to science attitudes, conspiracy theories, religion, environmentalism, and emotions related to climate change.

SKEPTICISM AS A CONTINUUM

According to the Yale Program on Climate Communication, only 72 percent of US adults believe that climate change is happening. Another 14 percent reject the veracity of climate change, and the remaining 14 percent do not know, are not sure, or decline to answer. Even fewer people believe that human activity plays a role in climate change: 57 percent believe in anthropogenic climate change, 30 percent do not believe that human activity impacts the climate, and 13 percent do not know, are not sure, or refuse to answer.[7] Here we consider what might happen if we were to combine the 14 percent of those who do not know if climate change is happening with the 13 percent who do not know if human action contributes to climate change to our conceptualization of climate change skepticism.

One may wonder why uncertainty or ambiguity about climate change should be considered as part of skepticism at all. We argue that in a time when information about climate change is prevalent and reliable data about climate change is easily accessible, being "unsure" as to whether or not climate change exists is, in fact, taking a position on this issue.[8] Therefore, uncertainty and agnosticism toward climate change should be considered in any analysis of climate change perceptions.

Our contention regarding the need to include those who are uncertain or uncommitted in analysis of skepticism is not original. Other scholars have made this same case. For example, Wouter Poortinga, Alexa Spance,

Lorraine Whitmarsh, Stuart Capstick, and Nick Pidgeon, in their article "Uncertain Climate: An Investigation into Public Skepticism about Anthropogenic Climate Change," suggest a need to include uncertainty in analysis of skepticism. They argue that "it is important to distinguish between different attitudinal terms such as skepticism, uncertainty and ambivalence."[9] Poortinga and his team presents uncertainty as an extension of skepticism, whereby skepticism can be understood as "a rejection of the tenants of mainstream climate science" and uncertainty as "a lower subjective sense of conviction or validity as to whether climate change 'really' exists, is caused by human activity, and/or will have major impacts."[10] They argue that uncertainty regarding climate change skepticism among the public is caused by the false perception that there is not a scientific consensus on the issue, which they attribute to media stories that present climate change as controversial alongside a powerful climate change denial countermovement.[11]

For our analysis here we use the continuum markers we developed in the same article in *Global Environmental Change*. Each represents a specific point on the climate change skepticism continuum:

EPISTEMIC DENIAL: People near this end of the skepticism continuum (whom we call epistemic deniers, for clarity) reject the idea that climate change is occurring and do not believe it is impacted by human activity.[12]

EPISTEMIC DOUBT: This is the middle of the skepticism continuum. People near this point (epistemic doubters) are both unsure if climate change is occurring and unsure if human activity impacts the climate.

ATTRIBUTION DENIAL: This too is in the middle of the continuum. People who fall near this point (attribution deniers) are those who believe that climate is changing but perceive the change to be a natural phenomenon not linked to human activity.[13]

ATTRIBUTION DOUBT: This is the other end of the skepticism continuum. People near this point (attribution doubters) are those who believe that climate change is happening but are unsure if the phenomenon is human caused.

Epistemic Denial

Blake is what many nonskeptics imagine when they envision a stereotypical climate change skeptic: a white man in his sixties with strong opinions and a tendency to discuss extreme politics that, to many, seem off topic but ultimately prove connected to his thoughts on climate change. Growing up with ideologically conservative parents, Blake spent time in both Seattle and rural Washington. Now he lives in rural Idaho and calls himself a "philosophical anarchist" who sees value in certain governance structures but believes the state overextends its "powers." During his interview Blake brings up Nazis and Nazism with surprising frequency. He casually mentions them when he discusses riding on a school bus to school, "like every good Nazi German kid." He also refers to Nazism when he discusses his belief that the United Nations is manufacturing the threat of climate change to create international cooperation. Blake contends that

> if you have, say, Germany start to militarize again, immediately everyone would be working together. In the absence of threat, you see disintegration of global coordination. So, this was a known and perceived need. So, how do you get all of the countries working together toward a common goal? How do you instill a sustainability method across nations? We needed some sort of global threat. So that was stated. That is not my conspiracy theory.

Blake's individual beliefs about climate change need to be contextualized, as they relate to others near the epistemic denial point on the skepticism continuum. Epistemic deniers make up approximately 18 percent of our survey participants. Like Blake, other epistemic deniers in our sample tend to be older (48 percent are above sixty years old) men (62 percent). People in this category tend to be more religious and more politically conservative than their skeptic peers. Three in four also believe that climate change constitutes a hoax, indicating that epistemic deniers are the most likely group to adhere to conspiracy theories related to climate change. This is not surprising, given that they are the group most likely to be exposed to climate news disseminated via the conservative echo chamber. For example, epistemic deniers express the highest reliance on *Fox News* for climate information and the lowest trust in other media, compared to other skeptics along the continuum. (For more details about these demographic and contextual variables, including percentage/mean distributions and significant test results, see the appendix, table 2.)

Blake believes that climate change is manufactured to promote international peace. He extends this argument by further suggesting that the choice of climate change (as opposed to another socially coalescing issue) was in response to increased interest in Middle Eastern oil. He contends that European governments felt economically threatened by an associated strengthening of Middle Eastern economies and therefore came together through the United Nations to manufacture a myth to undermine the oil industry. This collusion was advanced by "elitists," including Al Gore, who Blake views as by "every definition an elite," joining forces with European nations and eventually coercing climate scientists to make an argument against the use of fossil fuels.

Other epistemic deniers in our sample share this perspective. Some 52 percent of epistemic deniers believe that Al Gore is involved in the climate change conspiracy, while 49 percent believe that the democratic party is behind the climate change hoax. Similarly, 48 percent blame "global elitists" and 40 percent cite the United Nations as involved in the promotion of a false narrative about climate change. When we asked epistemic deniers why they believe these entities are promoting a false story about climate change, 56 percent suggested it was "to gain power," 52 percent "to make money," 47 percent "to make average citizens easier to control," and 43 percent "to promote globalization."[14]

In addition to the more extreme view that climate change is a hoax (as opposed to a misinterpretation of data, for example), we find that epistemic deniers hold stronger views within several "types" of skepticism. In 2004 Rahmstorf developed a typology in which he offers three types of skepticism.[15] He outlines distinct beliefs regarding trend skepticism (whether or not climate change is happening), attribution skepticism (whether or not people are causing climate change), and impact skepticism (whether or not climate change will have a meaningful impact on the planet). Later work by McCright and Dunlap adds a fourth type: consensus skepticism (the disbelief that scientists agree on this issue).[16] Epistemic deniers are more likely than other groups to show evidence of all four—trend, attribution, impact, and consensus skepticism—regarding climate change.[17]

This is certainly the case for Blake, who rejects climate change based on historical periods of "global cooling" and "incomplete data." He also does not believe humans impact climate change, nor that climate change will negatively impact human populations or animal species. He rejects the perception that there is a scientific consensus on climate change by pointing to what he sees as undue pressure on scientists to conform to the climate crisis narrative. He argues that reports that indicate significant scientific agreement on climate change are "falsified" and he places the level of agreement closer to 57 percent as a result of a survey he read conducted by a graduate student.

Beyond showing greater evidence of trend, attribution, impact, and consensus skepticism, compared to others along the continuum, epistemic deniers are the most likely to use personal experiences and anecdotes to suggest that climate change is not a real issue.[18] Blake, for example, argues that he has seen similar levels of snowfall and snowpack over his lifetime. Though he does acknowledge that "my frame of reference personally is very short," he argues that personal experience is important when going against scientists who "deceive the population on purpose or on accident."

In addition to questioning the scientific consensus around anthropogenic climate change, people near the epistemic denial point on the continuum have higher levels of concern than their peers about the methodologies used in climate science, the trustworthiness and motives of climate scientists, and the perceived undue influences of incentive structures in climate science.[19]

Blake, for example, believes that climate scientists operate in a "top-down system, where they are being told essentially what to do" and are prohibited from receiving funding if they do not affirm the climate change story. In Blake's words, "There's top-down funding for research and, if you don't agree with, if you're not doing climate change as we've defined it in research, then it's harder to get funded. If you're outspoken, if you're a really vocal scientist outspoken against climate change . . . your career is done." Regarding the broader climate conspiracy, he believes that climate scientists have been asked to backfill data to prove climate change exists, and "now we're trying to fill in details." He further believes that climate scientists do not have adequate data to make long-term claims about climate change, which undermines their claims: "If you don't have a baseline you can't say what is trending." Like Blake, nearly 70 percent of epistemic deniers in our sample agree that climate scientists "do not have enough data to know that human-caused climate change is happening" while 60 percent agree that "climate scientists are not open about their research." Further, 56 percent agree that "scientific journals only publish papers that conclude climate change is happening," while 77 percent affirm that climate scientists are influenced by funding.[20]

Our research further reveals that for environmental issues ranging from plastic in the ocean to sea levels rising, epistemic deniers demonstrate the lowest level of concern in comparison to other skeptics along the continuum.[21] This pattern also holds true in regard to pro-environmental policy support, where epistemic deniers are the least likely to support policies such as investment in solar and wind energy or regulating fuel efficiency standards for new vehicles.[22] Conversely, they are the most likely to support cutting funding for the EPA and withdrawing from the Paris Climate Treaty.

Blake does have some environmental concerns. He thinks pollution is a problem that needs to be managed. He is concerned about breathing dirty air. But his concern isn't particularly strong, as compared to others. He feels that the water systems are "pretty darn clean" and cleaner than they've been before. Blake argues that people should be concerned about pollution but not overly so and that there is a "threshold" where alarm begins to be warranted. According to Blake, we are not yet there. When we do get there, Blake theoretically believes that in state-level (as opposed to federal-level) regulations to control pollution will be warranted. Blake is not alone. Among epistemic deniers in our sample, only 40 percent expressed being concerned about air pollution while 46 percent are concerned about declining water quality, with 20 percent of them "not at all" supportive of federal regulation for controlling air pollution and 18 percent "not at all" supportive of the same for controlling water pollution.

It is important to note that, in general, these environmental beliefs correlate strongly with political conservativism. In our multivariate models we controlled for important demographic variables such as political ideology and gender and found that, compared to other skeptic groups, epistemic deniers still harbor lower levels of both environmental concern and pro-environmental policy support, but much higher levels of conspiracy ideation.[23]

Epistemic deniers are also the group least likely to express negative emotions such as worry, dread, sadness, or grief related to climate change.[24] Fear, and its variations such as worry and dread, are becoming increasingly more common responses to the topic of climate change.[25] Two-thirds of Americans now say they are at least "somewhat worried" about global warming, and one in four (26 percent) are "very worried" about it.[26] However, epistemic deniers consistently express less of these emotions in comparison to other skeptics along the continuum. Even when we controlled for a series of other theoretically pertinent predictors of climate beliefs (age, gender, education, income, religiosity, political ideology, race, and distrust in science), epistemic deniers expressed significantly lower degrees of worry, dread, and sadness compared to skeptics as a whole.

We asked Blake what he feels when he thinks about climate change. He responded by saying that when he thinks of the phenomena itself, he feels "no emotion." However, when he considers the propaganda machine that causes people in his life to believe in climate change, he feels "much the same way about friends who are being tricked into buying a house in a housing bubble," which we interpret as anger. This reaction makes logical sense. If someone does not believe in climate change, they will not have a strong reaction to discussion of the phenomenon. But, as Blake suggests, he does emotionally respond to what he perceives as deception on the part of global elites that harm people he knows.

To recap, people near the epistemic denial point on the climate change skepticism continuum, such as Blake, reflect what appears to be the extreme end for certain beliefs that correlate with skepticism. Epistemic deniers are the most likely to believe that climate change is a hoax. They are the least likely to trust science and scientists. They are the most likely to be steeped within the conservative media echo chamber. They hold the lowest levels of environmental concern and lowest support for pro-environmental policy. And they have the weakest emotional response to discussions of climate change.

Epistemic Doubt

Epistemic doubters make up 22.6 percent of our sample. They tend to be younger: over 40 percent of them are younger than age forty. They are also more likely to be women.[27] One epistemic doubter in our sample, Jill, is a college student. She grew up in Idaho and identifies as a Republican but recognizes that "there are good qualities on both sides." Jill was raised on a farm outside a very small town. On the issue of climate change, Jill says, "I'm not really like I completely disagree . . . this is something I don't know how I feel about . . . I'm still formulating my opinion." Throughout her interview Jill explains that she is uncertain whether or not climate change is happening, the role humans play in the climate crisis, and the potential impact of climate change on the planet.

One might posit, as we initially did, that people who do not hold an opinion on climate change are simply not paying attention. But Jill, and several other doubters in our sample, have read extensively on the issue. As a college student Jill even discusses the issue as part of her courses. While her college instructors largely operate from the foundation that climate change is real, caused by humans, and likely to be disastrous, Jill remains influenced by some of her high school teachers, who were less convinced.

Jill's perspective is complicated and shadowed by doubt. For example, when asked about increasing strength of hurricanes in recent years, Jill suggests that there isn't enough information to know if hurricanes are becoming stronger. She argues, "These things have been happening for so long, like just natural disasters in general." Jill believes it is possible that the earth's climate is changing but thinks that news reports are "blowing" news about climate change "up" to be bigger than it is. This causes her to feel as though assessing the truth on the issue is difficult. Jill doesn't believe that scientists are intentionally trying to mislead the public on the topic of climate change but argues that the information they present "is based on opinion." Jill's sentiments are shared by a portion of epistemic doubters: 10 percent of epistemic doubters agree that climate scientists "aren't doing real science," 23 percent agree that "doing science is the same as making educated guesses," 14 percent agree that "climate modeling isn't science," and only 22 percent explicitly agree that climate scientists "know what they are talking about."

Epistemic doubters are much less likely to believe climate change is a hoax than epistemic deniers and attribution deniers.[28] For instance, Jill's perspectives on climate scientists are distinct from those of Blake. Along with approximately 30 percent of epistemic doubters, Jill believes that the people advancing the narrative about climate change are generally trustworthy, but they simply don't have enough information to draw accurate conclusions and instead rely on opinion. As a result, Jill remains uncertain about climate change, falsely believing that climate scientists are similarly unsure.

Jill seeks more information about climate change. She wants to hear from the multitude of voices that "covers all spectrums" of this issue, and with more "facts" and fewer "opinions." She does contend that certain governmental organizations, such as the EPA and NASA, are more trustworthy than others because there are many people reviewing the work. As a result, she believes there is less opinion mixed in with data. These perspectives are represented among other epistemic doubters. For instance, while only 6.6 percent of epistemic doubters express "a great deal" of trust in media, respective numbers expressing "a great deal" of trust in the EPA, NOAA, and NASA are at 8.4 percent, 12.4 percent, and 17.7 percent.

Epistemic doubters have higher levels of environmental concern in comparison to the epistemic deniers. However, this level of environmental concern is lower than that of attribution deniers and attribution doubters. Similarly, epistemic doubters express higher levels of pro-environmental policy support than epistemic deniers, but again, lower than that of attribution deniers and attribution doubters. For example, while only 40 percent of epistemic deniers in our sample express concern about air pollution, 55 percent of epistemic doubters do, which is lower than that for attribution deniers and attribution doubters (approximately 80 percent each). Similarly, while 20 percent of epistemic deniers do not support federal regulation for air pollution, only 13 percent of epistemic doubters do, which is higher than that for attribution deniers (5 percent) and attribution doubters (3.8 percent). When all other skeptics are treated as a single group, epistemic doubters have lower levels of environmental concern and policy support by comparison, although these levels are still higher than epistemic deniers who are on one end of the skepticism continuum.[29]

This relative position of epistemic doubters' beliefs is seen in Jill's comments about the environment; she believes people need to have a "balance." Jill suggests that "you don't just go throwing your garbage out the window or anything" but also that there shouldn't be regulations placed on pollution. She argues that the government should instead "be reminding" people to dispose of waste properly and "encourage" them to make "smart choices," but that it "just comes down to people making the choice." Overall, 13 percent of epistemic doubters are "not at all" supportive of federal regulation for controlling air pollution, while 10 percent of them are "not at all" supportive of the same for controlling water pollution.

When we consider emotional responses to climate change, epistemic doubters experience more worry, dread, sadness, and grief compared to epistemic deniers and attribution deniers, but less than what attribution doubters experience.[30] For example, the average degree of worry related to climate change expressed by epistemic doubters (2.75 on a 1–7 scale ranging from "not at all" to an "extreme amount") is higher than that of epistemic deniers (1.99) yet lower than that of attribution doubters (3.38). Similarly, epistemic doubters report more sadness (2.72) than epistemic deniers (2.08), which is higher than that of attribution deniers (2.55) but lower than for attribution doubters (3.26). When other important predictors of climate beliefs are held constant, epistemic doubters express significantly lower levels of worry and dread compared to all other skeptics, yet the magnitude of the difference isn't as high as that for epistemic deniers versus all other skeptics.[31]

Jill says that she "doesn't know how I feel about it." When pushed further, she indicates that she feels "kind of a little bit of confusion, probably." Her reaction reflects a significant departure from Blake's, who claims to feel nothing but also indicates wavering uncertainty about the issue.

All in all, epistemic doubters are less likely than epistemic deniers to believe that climate change is a hoax. They are also less likely to distrust science and scientists and have higher levels of environmental concern. Finally, epistemic doubters have stronger emotional responses to climate change than do epistemic deniers. However, as we will show, epistemic doubters do not fall neatly in line next to epistemic deniers on the continuum of skeptic beliefs. Rather, they and attribution deniers interact in complicated ways to fill in the middle portion of the skepticism continuum.

Attribution Denial

Nearly 18 percent of our respondents are attribution deniers. Compared to attribution doubters and epistemic doubters they are more likely to be older, politically conservative, religious men and are similar to epistemic deniers for these demographic factors.[32]

One attribution denier in our sample, Douglas, grew up in a family of teachers and, as an adult, became a high school biology teacher himself. Originally from Appalachia, he moved to Idaho as an adult. He has been active in local environmental groups and has particular concerns about local biodiversity and bird habitats. He left the environmental movement recently because he feels like "climate change hogs the environmental debate." His political opinions have waxed and waned over his life, but he currently identifies as a conservative, much like attribution deniers in general.

Our survey data and models suggest that attribution deniers constitute the middle of the skepticism continuum, along with epistemic doubters. On climate change beliefs regarding trends, human causes, negative impacts, and the scientific consensus, they express higher levels of skepticism compared to epistemic doubters, yet lower than epistemic deniers. Additionally, they are less likely than epistemic deniers and epistemic doubters to use personal experiences and anecdotes to justify their beliefs.[33] Consider that only 18 percent of attribution deniers agree with the statement "My personal experience tells me climate change is not real," while 67 percent of epistemic deniers agree with the same. Similarly, expressing impact skepticism, 63 percent of attribution deniers agree that the "potential negative effects of climate change have been exaggerated." This percentage is higher than that for epistemic doubters (35 percent) but lower than that for epistemic deniers (75 percent).

Like other attribution deniers, Douglas believes the climate is changing but does not believe that humans are causing the change. To draw this conclusion, he does not rely on his personal experience, but rather on a belief that scientists simply do not have enough data to draw conclusions regarding human impact on the environment. He says, "I think we have clear evidence that we do have a global mean temperature increase over the last seventy to one hundred years . . . but there are so many other [nonhuman] players in the system."

Personally, Douglas does not distrust science or scientists. Rather, he thinks science is used as a "weapon" in political battles. He distrusts the rhetoric politicians use to present scientific findings to the public and their attempts to "dupe" the public. Like Douglas, attribution deniers typically fall between epistemic deniers and attribution doubters in their beliefs related to climate science.[34] While they raise concerns about science similar to other skeptics, they are more likely to question scientific methods and incentive structures of climatology than do attribution doubters, who make the opposite end of the skepticism continuum. For instance, while 31 percent of attribution deniers agree that "climate modeling isn't science," this percentage is at 54 percent for epistemic deniers and 12 percent for attribution doubters. Similarly, 71 percent of attribution deniers agree that "climate scientists are influenced by funding," while 76 percent of epistemic deniers and 56 percent of attribution doubters agree with the same.

Douglas also does not have concerns about the potential impact of climate change. He feels that God is in control and will protect the planet. Despite his skepticism that people impact the climate, Douglas does think people should be good "stewards" of Earth: "I think that it's wise and good policy and stewardship to reduce emissions, regardless. If we are seeing manmade climate change or not. I'm not saying it's a moot point, but I'm saying we should be doing those things anyway."

Along the continuum of skepticism, attribution deniers have more concern about environmental issues than do epistemic deniers and epistemic doubters, but their level of concern is lower than that of attribution doubters.[35] Going back to the example of concern toward pollution, 14.8 percent of attribution deniers in our sample state that they are "very concerned" about air pollution. This percentage is higher than that for epistemic deniers (11.2 percent) but lower than that for attribution doubters (32.5 percent). Similarly, 18.2 percent of attribution deniers state that they are "very concerned" about declining water quality, which is a percentage higher than that for epistemic deniers (14.0 percent) but lower than that for attribution doubters (33.3 percent). Attribution deniers' environmental concern and pro-environmental policy support is not significantly different from other skeptics.[36]

Similarly, attribution deniers' experiences of emotions related to climate change, such as worry, dread, sadness, and grief, situate them somewhere in the middle of the skepticism continuum.[37] For example, the average degree of worry related to climate change as expressed by attribution deniers (2.57) is higher than epistemic deniers (1.99) yet lower than attribution doubters (3.38). Similarly, attribution deniers express more sadness (2.55) than epistemic deniers (2.08), which is lower than that for attribution doubters (3.26).

Douglas's emotional reaction to discussion of climate change is stronger than Blake's or Jill's. He reports that he is "pretty excitable and so [he] get[s] amped up" but that when someone tries to get him to "take a step of action" he "avoid[s] it." It appears he describes an initial emotional reaction that is then replaced by a sense of apathy. While Douglas does not specifically connect these points, it seems possible that his religious convictions or belief that humans do not impact climate change may drive this avoidance. In fact, behavioral scientists suggest that emotionally uncomfortable topics that are difficult to solve, such as climate change, often lead people to apathy. Some people respond to such situations, by metaphorically "sticking their head in the sand" in what has become known as the ostrich effect or information avoidance, especially when they sense they have little agency to effect change.[38]

The middle of the climate skepticism continuum is a bit murky. While epistemic doubters align more closely to epistemic deniers regarding their environmental concern, attribution deniers appear to neighbor them along other measures. Attribution deniers fall between epistemic deniers and epistemic doubters on levels of political conservativism, the belief that climate change is a hoax, religiosity, strength of doubt by "type," trust in science, and emotional response to climate change.

Attribution Doubt

Attribution doubters make up the largest category in our sample, at 34.2 percent of respondents.[39] When compared to others on the continuum they most closely mirror epistemic doubters. They are younger (36 percent are younger than forty), more likely to be women (55.3 percent women vs. 44.2 percent men), less religious, and less conservative than epistemic deniers or epistemic doubters. Only 29 percent of attribution doubters are conservative, as compared to 62.6 percent of epistemic deniers and 32.3 of epistemic doubters. They are the group least likely to believe that climate change is a hoax (only 2 percent affirmed this) and the least likely to rely on conservative media such as the Fox Network for their climate change information. They also show higher levels of trust in mainstream media compared to people in other skeptic categories.[40]

One attribution doubter in our sample is Savannah. Savannah is deeply concerned about ocean wildlife. When she was a teenager her family visited the Oregon coast and it spurred a lifelong passion about ocean health, ice cap melting, and protection of arctic ecosystems. Savannah grew up in Idaho, moving around a bit, and has been rather isolated. She is from a conservative fringe religious group and was homeschooled; her social activities as a youth were largely restricted. Savannah identifies as politically "on the line." She struggles with liberal perspectives on abortion and conservative positions on the environment.

Evidence from our surveys show that attribution doubters constitute one end of the skepticism continuum and represent the group least likely to harbor the skepticism types identified in prior literature (trend skepticism, attribution skepticism, impact skepticism, and consensus skepticism).[41] As examples, consider that in our sample only 33 percent of attribution doubters agree that "potential negative effects of climate change have been exaggerated," compared to 74 percent of epistemic deniers, 35 percent of epistemic doubters, and 63 percent of attribution deniers. Further, while 50 percent of attribution doubters agree that "climate change is just natural variation," more epistemic deniers (73 percent) and attribution deniers (80 percent) agree with this statement.

Savannah believes that the climate is changing. Though she does believe that humans can negatively affect the environment, and is particularly concerned about Arctic and ocean health, she thinks climate change is "natural":

> It is very true that there is a natural climate change that happens naturally in the world. The way that our earth moves. The way that, there's like a, I want to say a climate season, a climate season, I guess, is the way I would put it. If you go across back in our history, you'll see that the earth has almost a clockworklike seasonal change, where it will go, like you'll have an Ice Age and then you'll have a warm[ing]. You'll have it cool off and then warm up and then you'll have an Ice Age and the world just naturally does do this. So that is true.

From this statement one might argue that Savannah is an attribution denier, but she goes on to argue that humans might have an impact with "our factories, our cars, and the amount of industrial stuff that we are doing." In short, Savannah believes "there's so much more we could know."

What Savannah is certain about, however, is that the earth will survive any change. She believes that the earth can and will "heal itself." However, she does have concerns about certain animal populations that will be lost because of climate change, whether it is human caused or not. She also fears an increase in extreme weather events as a result of the changing climate. Like Savannah, many attribution doubters also worry about extreme weather events: approximately 20 percent say that they are "very concerned" about heat waves (compared to 5 percent of epistemic deniers), and 38 percent say that they are very concerned about stronger forest fires (19 percent for epistemic deniers). Furthermore, 32.5 percent state that they are "very concerned" about animal species extinction (15 percent for epistemic deniers).

On average, attribution doubters are the least likely to question the science around climate change and the motivations that drive climate science and scientists.[42] Savannah believes that scientists are generally trustworthy and does not see them as maliciously motivated or coerced like many of her peers do. However, she does worry that scientists simply don't have all the information they need to draw accurate conclusions regarding climate change, which leaves her uncertain about the degree to which human behavior impacts the climate. In fact, 2.9 percent of attribution doubters "strongly agree" that "climate scientists do not have enough data to know that human-caused climate change is happening." This number is much higher for epistemic deniers (28 percent) and attribution deniers (15 percent).

Among all skeptic types, attribution doubters exhibit the highest levels of environmental concern and support for pro-environmental policies.[43] At her core Savannah is an environmentalist who is unsure as to whether or not humans are causing climate change. She fears for animal population loss and the destruction of the oceans. In fact, 29 percent of attribution doubters state that they are "very concerned" about habitat destruction and 32.5 percent state the same about animal species extinction.

Like Savannah, attribution doubters express the highest levels of worry, dread, sadness, and grief related to climate change compared to the other three groups of skeptics.[44] Even when other theoretically driven climate belief predictors such as political ideology and gender are controlled, our regression models reveal that attribution doubters experience higher degrees of these four emotions in comparison to all other skeptics. For example, the average degree of worry related to climate change expressed by attribution doubters (3.38) is higher than that of the three other groups: epistemic deniers (1.99), attribution deniers (2.57), and epistemic doubters (2.75). Similarly, attribution doubters express more sadness (3.26) than epistemic deniers (2.08), attribution deniers (2.55), and epistemic doubters (2.72).

Savannah's primary emotion with regard to climate change is grief. She fears the loss of animal species and habitats and imagines that "it just would be a really sad world for my children not to be able to experience that." When asked what she feels about climate change, she says "concern . . . sadness . . . but mostly sadness." Yet she stresses that she moderates these feelings by forcing herself to be "patient" in light of her belief that Mother Nature will heal herself and not allow the destruction of the planet.

Some may reject the inclusion of Savannah in our analysis of climate change skeptics. We argue that her perspective, and those of others similar to her, needs to be included and examined for several reasons. First, she identifies as skeptical about climate change and self-selected into a study to discuss climate change skepticism. Second, her perspective adds important nuance and understanding to perspectives on climate change—she believes it is happening, doesn't know if humans are causing it, thinks the earth will heal, and is an ardent environmentalist. By understanding the perspective of attribution doubters, scholars may find key insights into climate communications to help scientists and politicians better express to skeptics their findings about climate change. Finally, if we consider perspectives on climate change as a continuum or spectrum, people like Savannah help us understand the middle of the spectrum, where skeptics meet accepters.

Attribution doubters, like Savannah, make up the second end of the skepticism continuum as we have presented here. However, it may be possible to extend the spectrum further, using Rahmstorf's typology and by consideration of impact skepticism and doubt. We find that attribution doubters are, compared to others on the continuum, the least politically conservative, the least likely to believe that climate change is a hoax, the most likely to trust scientists, the most likely to have environmental concerns and support environmental policy, and have the strongest emotional reactions to climate change.

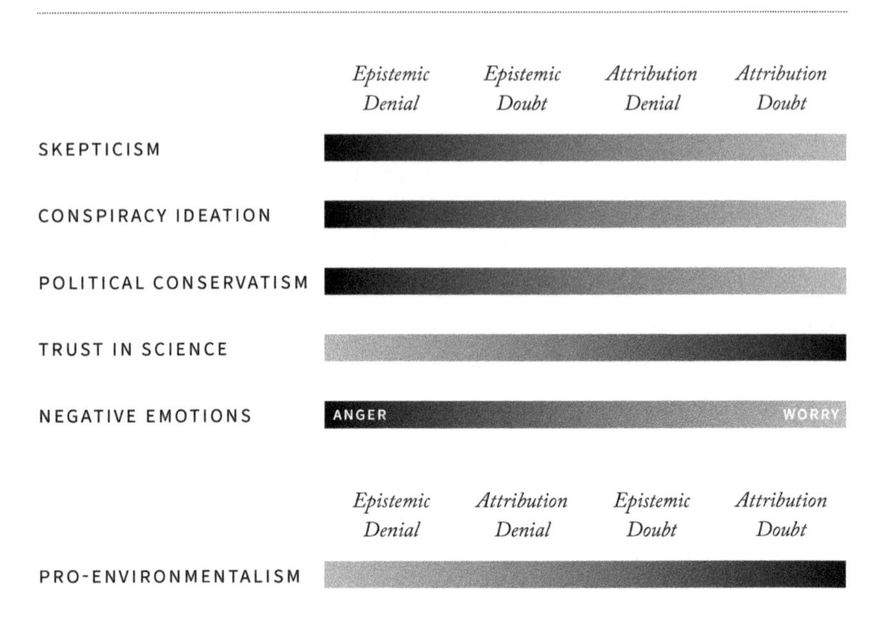

NUANCED PERSPECTIVES

Not all climate change skepticism is the same. By including those who are uncertain about climate change in our analysis, we find important distinctions in the perspectives of climate change skeptics along the continuum of skepticism. The figure above helps visualize the important points along the continuum in conjunction with skeptical beliefs on science attitudes, conspiracies, and environmentalism.

Epistemic deniers like Blake—those who outright reject that climate change is occurring—consistently make up one end of the climate skepticism spectrum. They are the most politically conservative. They are more likely than other skeptics to believe that climate change is a hoax. They are the most religious. They have the highest level of trend, attribution, impact, and consensus skepticism. They are the most likely to use anecdotes to reject climate change. They have the lowest level of trust in scientists and mainstream media. They have the lowest level of environmental concern and support for pro-environmental policies. Finally, they experience the lowest level of negative emotions like worry, dread, sadness, and grief over climate change.

At the other end of the continuum are attribution doubters like Savannah—those who are unsure if climate change is caused by humans. Compared to other skeptics, people in this group are the least politically conservative and the least likely to believe that climate change is a hoax. Attribution doubters are also less likely than other skeptics to express trend, attribution, impact, and consensus skepticism. They are the least likely to use anecdotes to reject climate science and they have the highest level of trust in scientists and mainstream media. They also have the greatest amount of concern for the environment and support for pro-environmental policies. Among skeptics, they experience the highest levels of worry, dread, sadness, and grief over climate change.

One exception to their otherwise consistent place on the skepticism continuum is attribution doubters' religious beliefs. We find that they are slightly more religious than epistemic doubters but that attribution doubters in general are less religious than deniers. For instance, in our sample of skeptics, 36 percent of deniers "never" attend religious services, in comparison to 44 percent of doubters. Additionally, only 13 percent of doubters attend religious services every week, while 22 percent of deniers do the same.

Epistemic doubters and attribution deniers do not extend the continuum uniformly. Rather, they add important nuance to the middle of the skepticism continuum. On specific climate beliefs, for example, epistemic doubters are less likely than attribution deniers to hold trend skepticism, attribution skepticism, impact skepticism, or consensus skepticism. On emotions, epistemic doubters are more likely than attribution deniers to experience worry, dread, sadness, and grief over climate change. Further, epistemic doubters are more likely than attribution deniers to believe that personal experiences and anecdotes prove climate change is not occurring.[45] They also have lower levels of trust in science than attribution deniers and doubters. In this regard it appears the strength of a skeptic's convictions (denial vs. doubt) has the strongest influence on perspectives related to climate change and science.

In contrast, on issues of environmentalism, attribution deniers have higher levels of environmental concern and support for pro-environmental policies than do epistemic doubters and deniers. It appears, then, that the specific issue being questioned (natural climate change vs. human impact) influences skeptics' environmental perspectives more than the strength of their convictions (denial vs. doubt). In other words, people who deny that humans contribute to climate change actually hold more pro-environmental views than those who doubt that climate change is occurring.

While our analysis reveals important nuances along four critical points of the skepticism continuum—epistemic denial, attribution denial, epistemic doubt, and attribution doubt—when comparing overall denial versus doubt, we also uncovered some important differences. Overall, our survey sample consisted of 38.5 percent deniers and 61.5 percent doubters. Compared to doubters, deniers express more of all four types of climate change skepticism: trend, attribution, impact, and consensus.[46] Deniers are also more likely to use personal experiences and anecdotes to question climate change.[47] Conversely, deniers overall are significantly more likely to believe that climate change is a hoax (89 percent), compared to doubters overall (11 percent).[48] In terms of their emotional responses to climate change, deniers express significantly lower worry than doubters (2.28 vs. 3.13), a trend that is repeated across dread (2.07 vs. 2.71), sadness (2.31 vs. 3.05), and grief (1.99 vs. 2.52). These overall comparisons between doubters and deniers affirm our suggestion earlier: the strength of a skeptic's conviction (denial versus doubt) matters and significantly alters how a skeptic views and emotionally responds to climate change and associated environmental issues.

We also find other interesting and important pieces regarding the climate change skepticism puzzle. Women are significantly more likely to be doubters (64 percent) of either type than deniers (36 percent). Men do not seem to show any sort of trend regarding the strength of their convictions. This may be reflective of a broader cultural trend in which women's confidence is undermined. In their book *Womenomics*, reporters Claire Shipman and Katty Kay find that even among the most objectively successful women, self-doubt causes women's confidence to waiver.[49] We suspect that the cultural forces that lead women to question their opinions and capabilities are reflected in their increased representation among doubt-related points on the skepticism continuum.

CONCLUSIONS AND IMPLICATIONS

We have explored the relationship between the strength of climate change skepticism, on a scale of uncertain to certain, as it associates with other variables. We find that the strength of one's skepticism correlates with certain demographic factors, level of environmentalism, trust in science, degree of influence from conspiracy theories and religious narratives regarding climate change, media and information access, and emotional reactions to climate change. We also find interesting nuances in the middle of the skepticism continuum, wherein attribution deniers align more with epistemic deniers on the nature of their beliefs about climate change, their use of anecdotes to dismiss climate science, and their trust in science overall. Yet we also find that epistemic doubters align more with epistemic deniers regarding their environmental concerns and support for pro-environmental policies. These findings suggest that not all climate change skepticism is the same and that important nuances exist along the continuum of skepticism.

Early on we focused on the ways that climate change skepticism operates as an identity for skeptics. Now we point to meaningful variance within this broad identity category. This delineation may reflect what psychologists call salience—or the degree to which an identity is primed or central to an individual. For example, sociologists Jan Stets and Peter Burke discuss the influence of identity salience using the example of a hypothetical scientist. They explain that those who enact the behaviors associated with science best are likely to be more successful in their work. It may also be the case that, for these scientists, their scientist identity is more salient.

When we look at the case of climate change skeptics, our interview data suggests that those who hold strong beliefs that climate change is an orchestrated hoax (epistemic deniers) have a pronounced sense of being marginalized. They may have the strongest sense of persecution by climate scientists (or at least the greatest distrust of them). Perhaps as a result of the clearer boundary that results between group memberships, the people who best enact the behaviors associated with an identity are often held up as examples of group membership.[50] As a result of this—and media pressure to tell a good story—the celebrated climate change skeptic tends to be on the extreme end of the skepticism continuum.

Among other implications, our findings here suggest the need to include people who hold uncertainty about climate change in conceptualizations of climate change skepticism. Ignoring those who say they are "not sure" about climate change or humanity's role in climate change omits important perspectives, such as those of Jill and Savannah. Jill and other epistemic doubters clearly hold important views in the understanding of trend skepticism. Missing out on these important perspectives means potential gaps in our understanding of tailored communications and areas of common ground for policy support.

We see our work, further, as extending Dunlap's continuum, to range from those who remain unsure (who have been exposed to information about climate science and refuse to make a conclusion on the issue) to skeptics (who express a position but remain open to information) to deniers (who refuse to accept new information and have closed their minds off from possible change).[51] We suspect that our continuum could conceivably be extended to include those who believe climate change is happening and caused by humans but reject or are unsure about its impacts on humanity or the earth (impact skepticism and uncertainty). Therefore, in our research we operationalize skeptics in opposition to believers. Our work includes both those who are active in their skepticism, the epistemic deniers and attribution deniers, but also those who are unsure, the epistemic doubters and attribution doubters.

It seems that Nick's perspective on the variance of climate change skepticism is most relevant: not all skeptics are the same. Often in media stories about climate change skepticism people like Blake and Nick are featured as quintessential skeptics. This image distorts the reality of skepticism and exacerbates already existing ideological divides. It makes collaboration and communication difficult. Our findings suggest that many skeptics are not as convinced in their positions as nonskeptics often believe them to be. Rather, people like Savannah and Douglas may be interested in collaboration with nonskeptics on certain environmental issues and people like Jill and Savannah may be willing to engage in meaningful discussion about the veracity of climate science. By understanding the nuances of skepticism we can improve communication between skeptics and nonskeptics and emerge from these discussions with shared visions for environmental healing.

eight

CHANGING PERCEPTIONS OF CLIMATE CHANGE

OVER THE LAST DECADE, rates of belief in climate change among American adults have slowly but consistently increased. In 2010, 57 percent of Americans believed the climate was changing, 48 percent accepted that it was human caused, and only 33 percent thought there was scientific consensus on the issue. By 2021, 72 percent of Americans believed climate change is occurring, 57 percent accepted that it is human caused, and 57 percent recognized the scientific consensus on the issue.[1] However, severe political partisanship remains, with 86 percent of liberal Democrats and 75 percent of conservative or moderate Democrats accepting the veracity of anthropogenic climate change, compared to only 27 percent of conservative Republicans and 48 percent of liberal or moderate Republicans who accept the same.[2]

Recent work by political scientist Matthew Motta, published in the *International Journal of Public Opinion Research,* uses data from three survey-based studies (Pew Research, Cooperative Congressional Election Study, and Yale Program on Climate Change Communication) to show that this slow change in perspective does not reflect the changing of hearts and minds of the American public but, instead, primarily demonstrates an increase in political progressivism among millennials as they age.[3] Motta calls this the "changing samples model," that a change in perceptions of climate change is actually an increase in ideological liberalism and educational attainment among young adults.[4]

In truth it is very rare that individuals shift their thinking on deeply partisan topics or change their membership in a political party. To garner greater acceptance of climate change and associated support for policy initiatives to mitigate its threat, we need to understand what happens in the uncommon cases of people who do change their minds about politically divisive issues. Stated simply, we do not have time to wait for political will to swing in support of drastic climate policy.[5]

Much of the scholarship on how people change their minds on climate change is theoretical, suggesting that certain types of media frames may be most effective or that people's minds might change if they encounter more information on the subject.[6] But it is difficult to find a large sample of people who have changed their minds about climate change because they simply do not exist.

In this chapter we present preliminary findings from our second interview project, among people who have shifted their thinking on this issue, and builds off our work published in *Rural Sociology*.[7] To contextualize their experiences, we draw on ideas on identity, ideology, and media choice and delve into the processes people undergo when changing their minds about climate change.

Our findings reveal two different trajectories of change for skeptics, depending on the nature of their skepticism. For people who move from actively denying climate change (i.e., deniers), shifting beliefs about climate change occur via a profound need to reconcile what emerges as cognitive dissonance when their underlying religious identities are challenged. For skeptics who move from being unsure about climate change (i.e., doubters), an acceptance of climate science happens when they either encounter new information from a trusted source or personally observe the effects of climate change.

CHANGING POSITIONS ON POLITICAL ISSUES

Scholarship on the topic of changing minds on politically partisan issues has been complex and varied in its approach. A great deal of the work has centered on the role of media and message framing as they shape popular perspectives. For example, Nisbet looks extensively at the topic of framing news stories about climate change and how telling a narrative in a particular way affects the perceptions of those who encounter it. Nisbet argues that certain frames, or ways of telling the story of climate change, may reinforce political partisanship on this issue.

Others approach the discussion through the lens of shifting identities. Work in psychology demonstrates the power of identity in shaping people's behaviors, emotions, and worldviews. The development of new group identities can thus alter perceptions and experiences. Psychologist Salma Mousa examines the theory that the creation of new group identities can override out-group derogation. In 2016 she went to northern Iraq and created a soccer league of mixed religious teams, combining populations of Muslim and Christian players. Her work demonstrates that through the creation of a shared identity as members of the same team, and subsequently sharing a set of goals, previous in-group/out-group dynamics could be overridden.[8]

Yet others see such ideological changes as reflective of increasing scientific literacy, the so-called information deficit model. This model demonstrates that in situations in which a gap exists between scientific knowledge of an issue and popular opinion on that topic, the divide is based on a lack of shared information. The solution to overcoming the divide is a unidirectional communication strategy whereby scientists share their knowledge with the public. This model has been used to drive substantial media campaigns to advance climate change science through social media, news media, documentaries, and other sources. However, this model has been critiqued as both "overly simplistic" and "inaccurately characterizing the relationship between knowledge, attitudes, beliefs, and behaviors."[9]

In the following analysis we explore the pathways people traveled as they changed their minds on climate change, beginning with people who originally denied climate change and ultimately came to accept climate science.

We then follow the trajectories of people who were originally doubters but ultimately became believers. We identify and build on insights from identity theory, cognitive dissonance, information deficit model, and media studies.

FROM DENIER TO ACCEPTER: IDENTITY THEORY AND COGNITIVE DISSONANCE

The experiences of participants who shifted their thinking about climate change reveals a complex unraveling of a web of identities and ideologies. It further unearths the difficult work required of participants to resolve cognitive dissonance as they attempted to reconcile the early political identities and ideologies to which they adhered with new information they learned about the world, about themselves, or about people they love.

This unraveling is not a simple one-time event; it happens as a constellation of coalesced events where people who denied climate change and later accepted it did so gradually following significant changes in their lives. They moved from home to attend college. They left old peers who shared their conservative values and met new people with a more progressive mindset and diverse experiences. They encountered classrooms where critical thinking, an open mind, and willingness to learn were key to success. They had their existing ways of thinking challenged in small fissures until they were more widely shaken apart.

Six participants moved from denying climate change to accepting climate science. For three of them the shift happened through a process of questioning the hard-handed religious ideologies in which they had been raised, moving to a new location, leaving a peer group with a shared identity, and encountering pluralism in college. The three others had their religious ideologies challenged when they or people they loved came out as queer. In all six cases the cognitive dissonance that emerged resulted in a significant shift in worldview, including their perspectives on climate change and their identities as climate change skeptics.

Changing Social Locations and Going to College

Though the sample for this pilot project is small, we find several key themes that can serve as a foundation for future work on this topic. As mentioned, all six deniers changed their minds while in college. It is difficult to tease apart the mechanisms of influence that leads people to change their minds at this time in their lives. From our interviews and extant research, it appears that much of it is an outcome of changing peer groups because of a significant geographical move; having the opportunity to learn about science in a space in which doing so is socially supported; or, for some, having religious ideas challenged by the emergence of a new, conflicting identity—sexuality.

Scholarship on identity change informs us that shifts in one's identit(ies) is an incremental and slow process.[10] Changes typically occur when people have difficulty finding "verification" for a particular identity. This can happen when people recognize a perceived inconsistency or hypocrisy present in an identity. Shifts in identity can also occur when an individual holds multiple identities that conflict with one another (identity conflict), such as seeing themselves as both evangelical Christians and gay, as is the case with some people in our sample.[11]

James's and Anna's experiences highlight the effect of identity conflict in changing beliefs about climate change. James and Anna are a married couple who participated in separate interviews with us. The two met and began a relationship in high school. Upon graduation, James joined the navy and Anna went to college. Two years later they married and bounced between locations due to his job, ultimately ending up concurrently attending school in Washington.

Both James and Anna grew up in conservative homes. Anna attended a religious public school, where she learned that "climate change wasn't real." Her ideas about several controversial topics—evolution, climate change, and abortion—were deeply intertwined. In her words, "Climate change is tied very closely to the ideas that I was raised with about evolution and about abortion." Beyond church itself, Anna reflects that her social community was made up predominantly of members of her church and that, among these peers, beliefs such as "Don't trust Al Gore, he's no good" were the norm.

Anna's ideas about climate change began to shift in college. She was in a new geographic location and found a church of a different Christian sect, one that was more open to scientific ideas. Anna believes that these conditions set her up for what became a pivotal moment in her thinking on climate change. Anna remembers a human anthropology class within which the professor talked about evolution. In this conversation the professor said, "Your religious beliefs don't have to conflict with evolution, that you could have both and they could coexist perfectly together." For Anna this was life changing: "That was just like this moment, like there can be both things . . . I think that was kind of the watershed moment . . . I can learn true facts and science and study these systems of knowledge that we have in place to understand the world and I can be a Christian and go to church and that's fine too."

James was also heavily involved in conservative Christian church groups growing up. Like Anna, his social group was "connected to my church so much, I tended to hang out with people from that." Among these peers James "heard a lot about the exaggeration of climate change," which aligned with his political values at the time: "I definitely branded myself a conservative in high school and growing up . . . it was very important to me to feel like I belonged in a group, just like everybody growing up as well." He continues: "That's a group of these high school conservatives that I felt really comfortable and wanted to be a part of . . . Whenever there was some sort of political message, that's how I would align myself. 'What did these people believe?' so I can be a part of their club."

James maintained these beliefs through his time in the military but began to question them once he moved across the country with Anna to attend college. He recalls:

> Just learning that scientists, that there wasn't a whole lot of debate whether climate change existed . . . Just understanding some of it, just the basic concepts of how science is done, I think was probably when that changed. It wasn't like a big light bulb moment, but . . . If you would've asked me a year before I went to college, I would have said, 'Yeah, climate change is exaggerated and we probably don't really have a part in it, and it doesn't matter even if we did because the planet would correct itself.' But maybe a year after I started going to college

> and . . . I definitely would've said that we had a part to play in it and climate change is happening at a faster rate than it ever has, and stuff like that.

It is difficult to separate the influence of shifting peer networks from scientific education in the change to our participants' attitudes about climate change in college. James explains this in his interview: "It was definitely a changing perception of science and I think that that probably went along with change in my political ideologies to a more left-leaning politics. I don't think necessarily that because I started hanging out with more liberal people that my views changed. I think that happens together."

As evidenced by Anna's and James's experiences, educational institutions play a central role in shifting people's perceptions of climate change. Colton's experience demonstrates this further. In his interview Colton spoke at length about how education was instrumental in shifting his thinking about anthropogenic climate change. Though Colton grew up in a progressive family and attended an ideologically aligned church, in high school Colton converted to a more conservative religious sect. Through this experience Colton said, he "kind of changed, or . . . kind of rejected a lot of [his] earlier beliefs." Colton became a creationist. He also accepted his religious peers' stances on climate change, that it did not exist. After high school Colton attended "a private religious university that was owned and operated by the church I had joined." While there, his beliefs about climate change were reinforced: "Some of the science classes that I took reinforced my climate skepticism . . . there were definitely messages like that that I heard that probably tended to make me more skeptical." For example, he tells of a story in which he "took a geology class . . . as part of the class we had to watch a video that was a documentary that involved some scientists who are climate skeptics . . . definitely made to make you think 'Oh, the scientific community is definitely not in a consensus on this issue.'"

It was not until graduate school, when Colton relocated to a new place to pursue a master's degree in environmental history and encountered a new social group as well as a novel, nonreligious-based curriculum, that his opinions began to change. In his graduate-level classes he "learned a lot more of the history of how Americans have shaped the environment here in the United States." He also "learned the term 'the Anthropocene' and the fact that geologists are now debating whether or not we're in a new geologic era that is created by human changes to the overall environment, not just the climate."

For course content and education to impact one's position on a highly politicized social issue, one needs to be open to learning new information. Our common psychological biases, such as confirmation bias—the tendency for people to seek out or accept information that aligns with their preexisting views and reject information that challenges them—acts against this. We see this bias operating prominently on issues that are politically divisive. For example, work by sociologists Jason Carmichael and Robert Brulle with environmental scientist Joanna Huxster finds that Republicans solidify their climate change skepticism in the face of stories confirming the veracity of climate change.[12] They dismiss new information and through the process of rejection hold tighter to their prior positions.

Yet something in the experiences of Anna, James, and Colton opened them up to new information. We suggest it is likely a combination of geographic relocation, associated shifts in peer groups, and the encounter with a culture in college more open to new ideas that together may have made their skeptic identity less salient and ultimately led to them abandoning it and its associated beliefs altogether.

Colton asserts that his experiences as a philosophy minor in his undergraduate education primed him to be open to learning new information as a graduate student. He reflects that "there were definitely a lot of things in my studies of philosophy that kind of opened my mind to new possibilities and challenged some of my preconceptions on what I believed and what I thought about politics, and religion, and those sorts of things." In graduate school, then, he "read a lot of books on environmental history." This experience of encountering new information changed Colton's mind on climate change. He considers "all these historians talking about all these environmental issues and the ways that we changed the environment, and oftentimes negative impacts that we had, it becomes kind of hard to be reading all of that and to still be like, 'Well, I don't think there's much to this.'"

Like Colton, Anna's ideas began to change when she attended college. Following that first anthropology class, Anna took several other courses that impacted her thinking about science and climate change more broadly.

She reflects: "My college experience and all the things I learned, even if it wasn't tied to the environment or climate change, made me understand what science was. That is what changed my mind." For Anna it was reconciling her faith with her scientific beliefs that ultimately led to her acceptance of climate science. In college she found a mainline Protestant church that was more "open to the ideas, the world." This empowered her to listen more openly to the lessons in her classes because "it wasn't in conflict anymore with the religious beliefs and identity that I had had around climate change. It wasn't in conflict anymore." In Anna's case it was not so much learning facts about climate change but being open to the concept of religious pluralism that empowered her to change her mind on the issue.

For Anna, James, and Colton, going to college, leaving their friends, and learning new information about climate change challenged their monist religious perspectives. In each case this new pluralist approach to thinking about issues, in partnership with finding themselves in a social location more open to the sharing of knowledge, enabled them to accept and learn from new information rather than turn to the more common tendency toward confirmation bias. This combination of factors served to shake their climate change skeptic identity.

Coming Out as a Gateway to Change

While changing peer groups, moving geographically, and having access to scientific education significantly influenced the participants who moved from denying climate change to accepting climate science, we also saw a second pattern emerging from our pilot dataset. Three other people in our sample had their own awareness of their sexuality and gender challenge their existing belief structures and ultimately unravel much of their conservative ideological perspectives.

Max, for example, grew up in northern Idaho in a very conservative household and town. His family had relocated there from California to escape "their lefty policies." He was raised in a conservative Protestant church, to which he was very committed. He first became aware of climate change listening to his "father yelling at the television or radio." While he talked to his dad intimately about political issues, for "six or eight" hours a day, he had a few friends with whom he shared political views and discussion.

Max's father and, by association and influence, Max himself, believed that climate change is a hoax that is orchestrated by

> liberals to try and get people to vote for measures that would strip their own power away from them and take away their individual freedoms, and it was something used by the government to push their power and bolster their reach, widen their reach, over the people . . . It was also used as an instrument of globalization, which we were very much opposed to, to integrate society when it didn't need to be integrated.

Things began to shift for Max in high school when his peer group matured. He recalls,

> Something weird happened with the friend group I was in. We banded together as children, but we didn't ever know why, we had nothing in common. Then, by the time we got to high school, probably by sophomore year, quite a few people in this group started coming out as LGBTQA and, by the time we graduated . . . eighty percent of the friend group was gay or bisexual.

Following their coming out, things changed with Max and his friends. Max explains: "With that came a splitting from the political views of the community, because that [being bisexual] was very much not okay within the community. So, they split away from that, and they got to look into what my father would call 'lefty culture,' and realized how much of it they liked."

Becoming aware of his friends' sexuality and changing political ideologies was difficult for Max because "at the time, since I was so deeply, deeply closeted that I didn't even realize it." Max felt significant cognitive dissonance in loving his friends but also holding ideological beliefs denying them rights and holding religious beliefs of their innate sinfulness. "But I still loved them because they were people and because they were my friends." A few years later Max "figured out" he was bisexual as well. This was pivotal for Max in his relationship to climate change and conservativism more broadly.

Max's new ideas about climate change occurred "really slowly . . . I was deeply invested for so long and just turning away from that, there's so much insecurity that I had with that." For Max, his sexuality was the catalyst for broader ideological change. He recalls: "My sexuality, that played a big spot in it, since one of the most grounded things that was holding me to those views was my religious views. Then, when you spring onto those religious views everyone in that religion hating you . . . You're trying to stay anchored with that but that which holds you is pushing you away . . . So, I was really lost."

This shifted further for Max during his senior year in high school. Max had a sociology teacher who taught students about "sociological imagination." Max wrote a paper for this class about his upbringing. He found support from this teacher and "that was one of the last things that broke my solid view of this conservative world that my father had helped construct in my mind." Having this compassionate teacher and learning to think critically about society "was one of the last things that really broke my solid hold on these views . . . I finally gave up Christianity and that was like the final bit." Max continues, "I'm a liberal now. Intersectional feminist, all that stuff."

For Max, a combination of recognizing that he was queer and encountering a teacher with whom he connected served to rattle his religious and political identities and beliefs. Having a close peer group of gay friends may have served as a counter in-group to Max, one in conflict with the beliefs and behaviors of his climate change skeptic peers. The resultant tension was resolved, for Max, when he met a kind and well-meaning educator who changed the way he viewed society. This combination of new information from a respected person and a social group with a counter-identity pushed Max to shed his incongruent skeptic and conservative identities. Max's experience reflects findings in psychology that demonstrate similar processes whereby membership in a new in-group can lead to new perspectives on the world and the rejection of previously held in-group/out-group beliefs and associated tendencies towards confirmation bias.[13]

Like Max, Alex's views on climate change coincided with their awareness of their gender identity and sexuality. Alex identifies as "nonbinary, more or less lesbian." Alex grew up in a military family and "moved around a lot," so they were homeschooled. Their key social network was "always a group of homeschoolers." Alex's classroom material was "from a conservative publisher" that downplayed the threat of climate change. In school, through their peer group, and from their parents, Alex consistently received the message, "It's not really happening. It's not a thing we need to worry about." They recall being taught that "scientists are atheists, and they are lying. Climate change is not a thing. It's not happening . . . It's one degree. What's one degree?"

Alex's perspectives on climate change and "on a lot of other things" changed "right after I graduated from college." Alex recalls: "That was when I started to figure out that I was queer. It was when I started to figure out that people lived different ways than I had lived and that people had different experiences." For Alex, much of this education occurred through reading personal memoir-like stories in online spaces. Here they encountered the experiences of other queer folks and began to shift in their thinking. "Figuring out that I was queer for sure solidified that a lot of things I had been taught growing up were maybe bunk." They continue: "It had never been explicitly said, but it was sort of implied, 'You can't possibly be queer, you're a Christian. All good Christians are straight. Sexuality is a choice' . . . And, so, I figured out that and I was like, 'Well, shucks, what else have they lied about?'"

For both Max and Alex, their experiences of recognizing their sexuality while among a conservative religious community that rejected them pulled the initial thread that began to unravel their broader political beliefs and, ultimately, their perspectives on climate change. These last comments by Alex—that the hypocrisy in faith led them to question all they had been taught—is particularly poignant.

Like the other participants in our sample, Samantha grew up in a conservative social environment. She attended public school and, though her parents were progressive, her peers were deeply conservative. Her ideas around climate change evolved from these peers and in opposition to her parents, "in an adversarial type of way." She explains: "When you're thirteen or fourteen and the only people saying it's one way are your parents and everybody else around you are saying it's the other way, you don't tend to believe your parents."

For Samantha, her views began to change in college when she became close friends with an out gay man. She had "never directly interacted with somebody who was gay at the time." She "was curious" but also "a little apprehensive." Samantha "talked to him a lot about his experiences and learned a little bit more acceptance." Shortly thereafter, "another friend of mine came out as transgender . . . and that further solidified that."

These friends (a new in-group) ended up guiding Samantha to new ideologies—she began to explore "the ideas of Bernie Sanders and democratic socialism and stuff like that." She reflects: "Talking to diverse people who have had experiences with things" has shifted her thinking about climate change. Samantha also came out as transgender several years after meeting these friends and, through this experience, became further connected to people across the globe. She explains: "Part of being trans is having a large internet support system. And so, I have a lot of friends online who are from various parts of the world." Samantha learned from her friends about climate disasters such as "flooding year-round" or "Zika" and "hearing first-hand evidence and seeing things and starting to think about it for myself."

The form of cognitive dissonance that Max, Alex, and Samantha experienced has been identified elsewhere in social science research. In 2012 a study led by Cindy L. Anderson reviewed existing research on the "religious identity/sexual orientation identity conflict."[14] This review demonstrates that the most common response queer Christians in conservative churches undertakes is to leave their church. Holding two identities in conflict leads to a need to reconcile the resulting cognitive dissonance. In the case of these three participants, this reconciliation was accomplished by rejecting their earlier identities as conservative Christians and/or political conservatives.

Reconciling Cognitive Dissonance

At the beginning we introduced climate change skepticism as a social identity. Social identities are those one holds in relation to members of a social group. In times when one encounters cognitive dissonance—such as when Anna learned that she could hold beliefs about evolution and at the same time believe in science—group membership can serve to reduce the need to resolve or reduce the sense of cognitive dissonance. However, in the absence of peers, people feel a more intense need to reconcile feelings of dissonance and they often do this by changing their minds about certain issues and reducing their sense of identification to their group. Said simply, when around other climate change skeptics, people may reject or ignore the sense of discomfort they experience when they encounter new information, but when lacking a peer social group, they instead begin to change their minds and alignment with skepticism.[15]

It appears that for our skeptics, moving to a new location and developing a new peer group were key to allowing them to reconcile their cognitive dissonance, changing their minds and reducing their identity as skeptics and conservative Christians. This is in line with social identity theory, which suggests that when a social identity is salient (or strong and meaningfully primed), people more readily align their beliefs with that of the group.[16] Had their skeptic friend networks still been present in their lives, research suggests, Anna, Colton, and James may not have shifted their thinking on climate change.[17] Instead, they may have continued the pattern of denying climate change and rejecting science to protect their social identity as conservative Christians.

While Anna, Colton, and James were faced with the challenge of reconciling competing ideas, Alex, Max, and Samantha had to make sense of holding two conflicting identities. The three identify as queer, and their religious beliefs suggested that being themselves is wrong. Research on the reconciliation of competing identities suggests that when people face tensions like this, they either change their religious beliefs to accommodate a new identity, leave their churches to accept their new identities, or try to live with dissonance.[18] According to research by sociologist Kimberly Mahaffy, one significant factor in determining which of these options an individual chooses is the age at which they develop a Christian identity and the age at which the individual realizes they are gay (in this study, lesbian). The older the age when they become a Christian, the less likely they are to leave the church or change their beliefs and instead continue to live with the dissonance. In contrast, the older a participant is when they begin to identify as a lesbian, the more likely they are to leave their faith or alter their beliefs to accommodate their sexuality.[19]

Identification with a group has profound impacts on sense of self, values, behaviors, emotions, and interpretation of events. The development of a group-based identity further serves to create a clear sense of in-group and out-group membership. Scholarship in psychology, then, questions what happens when new groups, groups that cross previous in-group and out-group divisions, are constructed. Psychologists Jay Van Bavel and Dominic Packer find that in such circumstances, people create new in-groups that can change their relationship to people previously perceived as others.[20] For our participants, going to college and developing new social circles with an interest in learning or coming out as gay or trans may serve to shift identities in such a way as to weaken previous in-group/out-group dynamics. This change subsequently opens people to new information and ideas they previously resisted or ignored.

FROM DOUBTER TO ACCEPTER

People who move from denying climate change to accepting climate science tend to have a slow but significant ideological shift. These shifts often are spurred or catalyzed by a change in geography, joining a new peer group, getting a college education, or realizing the sexuality of themselves or a loved one, all of which call into question the underlying religious ideas to which they previously adhered. In comparison, for people who move from doubting climate change to accepting climate science, two factors prove central: learning from people they view as respected authorities and witnessing aspects of climate change firsthand.

Henry came to accept climate science in college. He grew up in a rural part of New England, where climate change was rarely discussed and not a salient issue. As a result, Henry's perspective was "sheer ignorance." However, in college he had a professor who "was a big influence on how I viewed the issue." The class he took with the professor changed his worldview as it relates to climate change and his subsequent career trajectory. Henry credits that class as "really shifting kind of how, kind of my entire adult life . . . my career path . . . it really spurred an interest in science more than there already was."

For Henry it took a respected person in authority—an admired professor—for him to accept climate science. Another participant, Greg, also changed his mind following interactions with people he valued. Greg grew up in northern Idaho on a reservation. Like others in our sample, Greg's doubts about climate change stemmed from the narratives he heard growing up about other environmental problems. For example, he learned about things like "the whole ozone and acid rain" but he also saw how these problems were largely "resolved." This took some concern out of the threat of climate change.

Things changed for Greg not in college but when he began to hear about climate change from tribal elders. He explains: "Hearing it from your college professor is a different avenue than hearing it from your elders, because my authorities are older people." As Greg grew into adulthood he saw his tribe taking significant moves to combat climate change as best as possible. "[My tribe] had been here for so long that we know that climate change is a reality. It's been going on." In response, the tribe has created organizations to work on curbing, adapting to, and repairing the damage from climate change. In both Henry's and Greg's cases it was not only information that changed the participant's mind; it was that the new knowledge came from a source they respected.

While Greg and Henry were influenced by information provided from people they looked up to, other doubters shifted their thinking because of personal experiences with environmental changes. Consider Brent's experience. Brent grew up in Montana. Now retired, he spent his career in biological sciences. In his work he encountered stories about "global warming and Earth Day and stuff like that," but he resisted learning about the phenomena. He recalls: "I was never really on board with that. I didn't participate in Earth Day or things like that." Brent was resistant to things his town was doing to take environmental action. "I didn't see any relevance for the county to take action at that."

In 2010 Brent read an article about the impacts of climate change on sea level rise and the threat it posed to island nations. This primed him to pay attention to his local environment. He began to notice changes. Brent recalls buying a "secondhand weather station" for his son. They monitored the wind, which tended to be minimal. However, he "began to notice that—especially in the last couple of years—that we get a lot of wind at our house now. And there's a lot of swirling where the wind comes in and you can just see the trees go like this, and that never used to happen. That's a significant change." He also reflected upon his experiences with temperature. In the past Brent had "been able to cool our house by opening the windows upstairs, putting fans in the east facing windows, and then drawing air through the house . . . it would stay cool no matter how hot it got." But as time went on, Brent noticed the house was persistently hotter. He reflects that "it didn't cool at night." It impacted his well-being: "You can handle one or two nights like that—when you don't sleep well, and you go to work tired . . . but this endured, and it stayed so long that I . . .

had to break down and buy air conditioners for our windows. And, to me, that was significant. Since that time there has rarely been a summer when we have not put them in the windows and used them."

Like Brent, another participant, Eli, had his perspective on climate change shift based on his personal experiences with changing weather patterns. Eli grew up in a medium-sized city in the midwestern United States. His parents were both college graduates and his family attended a mainline Christian church in their town. His early views on climate change were largely shaped by other environmental issues, such as the hole in the ozone layer. He recalls that "the ozone had a hole in it and then it was healing. Things would come up, but they would get fixed relatively quickly." Thus, early on he presumed climate change was a nonissue that would get resolved.

Eli's views began to change when he became more aware of the long-term nature of climate change, especially when that awareness was rooted in his own direct experiences. "It became for me about seeing the damage that was happening right now was from practices done in the fifties and the Industrial Revolution." For Eli, "water scarcity" became central to his awareness of climate change. He reflects on his awareness of water shortages in the western United States but even how places like Wisconsin are "drilling deeper wells . . . and talking about building a pipeline from Michigan all the way across the state."

When Eli was a child he learned from his parents that "the world goes in cycles." They perceived that the climate might be changing but they didn't accept that "humans are the cause of it." Instead, they focused on stories such as "volcanoes erupting and that they produce way more CO_2 than we ever could in a year, just from blowing up." Through college Eli held onto these perspectives. It was not until he traveled to a tropical region in the Indian Ocean that his perspectives began to change. He recalls: "They have a lot of coral reefs there and because it was getting too hot, there was a lot of bleaching." After seeing this shift firsthand, Eli learned about similar bleaching in the Great Barrier Reef and elsewhere. It rattled him. "That is very unnerving to me."

Existing research on environmentalism indicates that personal experience with environmental disaster affects environmental views and behaviors. For example, researchers Wiwam Arp and Christopher Kenny find that people who live near toxic waste sites hold greater pro-environmental views than others.[21] This trend holds true for people who have been exposed to other environmental disasters, such as the Fukushima nuclear accident.[22] People who spend time outdoors as children, or who grow up discussing environmental issues and watching nature shows, also show higher levels of pro-environmentalism.[23] Furthermore, our earlier work shows higher levels of environmentalism among people skeptical about climate change if they have personally encountered negative environmental events.[24]

For people who are unsure about climate change, engaging with climate science with a person for whom they hold respect or having personal experiences with what they perceive to be a changing climate or changing environmental patterns appears to influence their ability to accept climate science. These preliminary findings warrant additional research but seem to be in line with existing work on access to information and the influence of personal experience on environmentalism.

Role of News Media

Although much research attention has been given to media framing and its role in changing minds, our pilot interviews suggest that the role this information plays is secondary or, perhaps, complementary to personal experiences. For example, Henry suggests that alongside his experiences in the classroom with his influential professor, he began to hear news stories about climate change. He recalls: "It was probably right around the same time when Al Gore was really leading the charge in his PR campaign" and that Gore's work, coupled with Henry's professor, was an important multiplying influence. Brent, too, says that when he began to witness changes in local weather patterns, the stories he heard on the news about climate change took on greater salience. He recalls hearing about sea level rise in Thailand around the same time that he noticed wind patterns changing in his hometown. This convergence magnified his awareness of climate change.

For others, once they began to change their views on climate change, media sources were used to confirm these novel perspectives. For example, after he began to accept climate science, Eli searched the news more for stories that confirmed his evolving thoughts. "I was just consuming more news or documentaries about it, but I realized that the problems weren't going to go away as easily, in a quick fix or anything."

People tend to seek out news sources that confirm or reflect the ideologies to which they already adhere. This results in what Kevin Arceneaux and Martin Johnson explain as "a public increasingly encouraged to only view the world from their point of view and, as a result, one that adopts more extreme and polarized political attitudes."[25] In effect, progressives and conservatives ultimately inhabit two ideologically different but "parallel realities."[26] It seems that for our participants, bridging that divide requires the impetus of personal relationships and experiences.

Media scholarship on climate communication offers many suggestions about how to best frame news stories about climate change. For example, environmental scientist Ezra Markowitz and psychologist Azim Shariff suggest that communication on climate change may be more effective if it focuses on emotions such as hope and empowerment and a sense of personal responsibility.[27] Nisbet's work builds on this framework by looking at the influence of how stories are told. In particular he presents what he calls the "public accountability" frame, which challenges what is often called the political right's "war on science" and argues that science is valid, it is about truth-seeking, and it is essential to listen to scientific findings even when the message is not desirable. Nisbet shows that this message galvanizes people on the political left while alienating the political right. Thus, he argues, we need to find shared values around which to frame news reporting on climate change.

Nisbet further highlights problems with framing climate change as an economic issue because there is too much uncertainty as to what is the best way forward. Instead, he points to several frames that seem more efficacious in rallying people across the political divide around climate change: framing the issue as one of moral or ethical responsibly or emphasizing the potential harm to public health. Yet, the prevailing COVID-19 pandemic and the polarization of public views about appropriate responses to it suggest that the specific topic under consideration also matters a good deal when determining the efficacy of frames. When an issue is heavily politicized, a public health frame may be less effective even though the target itself is a health crisis.

Other communication scholars suggest that the source of information may matter more than the actual knowledge shared. Carmichael, Brulle, and Huxster find that media reporting on climate change tends to result in confirmation bias and political polarization among those who read it.[28] They suggest that the source of information might be pivotal in shaping people's beliefs. Perhaps if the Fox Network reports on the veracity of climate change it might have greater influence on conservative thinking on the issue than if the same story is shared on MSNBC.

Our findings suggest that for skeptics the framing, the emotional content, and the media source matter less than the timing in which information is encountered. Our participants became open to new information about climate change either following a shift in social networks and location, or a change in identity, or upon personally observing environmental change. The presence of reliable news reporting on climate change matters. Encountering these stories in the news at the right time plays a role, but the details and framing of the stories appear less consequential. This is clearly an area ripe for additional experimental investigation.

Information Deficit Model

The participants who moved from doubting climate change to accepting climate science appear to align with, but also complicate, what scholars call the information deficit model. This model contends that one's failure to accept climate change is due to a lack of information or knowledge. They suggest that to fix this knowledge gap, information must be presented to the public to affirm climate science.[29] In spite of criticisms levied at the model, our work suggests that for people who doubt (but don't actively deny) climate change, this model may work, to some extent.[30] For Henry and Greg, especially, encountering information about climate change from sources they respected and viewed as authorities had an instrumental influence in their changing views on climate change. For others, such as Brent and Eli, media information on climate science served to supplement and contextualize their understandings of what they saw happening in their personal lives.

Yet for several participants it appears that access to information alone is not enough to alter their thinking on a topic such as climate change. Information needs to be presented by a respected source of authority or coupled with compelling, personal experiences. For doubters, this personal connection serves to reduce the psychological distance between the issue of climate change and one's own life.[31] In fact, studies have shown that personal experiences with certain types of extreme weather events increase belief in climate change, especially for people who are politically moderate or independent.[32]

CONCLUSIONS AND IMPLICATIONS

When we set out to study people who had changed their minds about climate change, we were unaware of how difficult finding such people might be. Indeed, people are highly unlikely to change their minds about politically polarizing issues. Yet our country's ability to effectively advance desperately needed climate policy hinges on the democratic process and garnering acceptance of climate science. The preliminary findings from our pilot interviews give some reason for hope and suggest specific ways that effective climate communication may progress.

We find that people who move from denying climate change to accepting climate science typically do so when their monist religious ideologies are challenged. For our participants this happened after people were primed to consider new ideas through a combination of moving to a new location, encountering new peer groups, and taking college courses in which the consideration of new ideas is a cultural norm. For others, this shift in religious identification occurred because of cognitive dissonance that arose when people came out. The incompatibility of their religious ideologies and sexual identities led them to question the antipluralist beliefs they had previously held, including those on climate change.

For participants who doubted climate change and later came to accept its veracity, the transition happened in a different way. Doubters came to accept climate science through personal relationships and conversations with people they admired and respected (professors or tribal elders) or through a timely merging of personal experiences with perceived environmental changes and awareness of climate change in the news.

In some respects, changing minds about climate change seems to be almost a magical aligning of events—either moving and finding new cultural norms or happening upon news stories that resonate with one's personal experience—or a radical identity shift that is impossible to replicate (i.e., coming out). However, these examples speak to the importance of cultural norms, values, and the availability of scientifically accurate stories about climate change and their implications on climate change communication and policy.

conclusion

MOVING FORWARD

AS WE WERE WRITING THE CONCLUSION for this book, explosive wildfires were raging in parts of North America; unprecedented flooding was occurring in China and India; and a powerful cyclone was causing destructive damage to the Philippines.[1] Climate predictions have gotten increasingly dire each year, with the 2021 IPCC report warning of imminent heatwaves, flooding, and droughts. That report moved UN Secretary General António Guterres to describe the situation as a "code red for humanity."[2] Fast forward to 2022: Guterres claims the world is now "sleepwalking to climate catastrophe." This, in light of the latest IPCC report of March 2022, which warns that effects of climate change "may soon outpace humanity's ability to adapt to it."[3]

Yet despite repeated and unprecedented weather disasters, the loss of human life, the visible changes to the landscape, the mass extinction of thousands of species, and the prediction of more to come, the international community, but especially the United States (the largest national contributor to elevated CO_2 levels per capita), lacks the political will to take action on climate change. Anthony Leiserowitz, director of the Yale Program on Climate Change Communication, contends that policies will not change in the absence of a public will, that is, demanded by citizens. For Leiserowitz, much of this challenge is rooted in the inflated attention and disproportionate power yielded to climate change skeptics.

As we stated at the beginning, the Yale Climate Opinion Maps (2021) show that 14 percent of the American public still does not believe that global warming is happening. Another 14 percent remain unsure.[4] And contrary to scientific evidence, a further 32 percent of the public believe that climate change is caused by natural changes, not human activities. The numbers get worse when we look at those who hold public office. The Center for American Progress reports that 139 Congress members (more than 25 percent in total) "deny or dodge clear scientific consensus, despite the obvious effects of climate change now accelerating across the country and globe." Exclusively, these skeptics are Republicans: 52 percent of House Republicans and 60 percent of Senate Republicans.

While it is easy to think of these skeptics as ignorant and obnoxious, or at best uninformed, they are, in fact, our leaders. But, more than that, to varying degrees they are our friends, family, coworkers, community members, and next-door neighbors. Rather than ignoring or dismissing their perspectives, we approached this project with open minds: to listen, to understand their concerns, and to formulate ideas to better engage with skeptics. To talk with each other, rather than talk past one another.

We hope that our efforts, in combination with the robust extant social scientific knowledge base on the topic of climate change, will help reduce tensions related to this topic and point to avenues for productive conversation. Ultimately we hope a nuanced understanding of skeptics' identities and associated perspectives, concerns, and thinking about climate change leads to the development of a comprehensive communication strategy targeting skeptics, which could generate wider support for climate policy and help mitigate climate change.

With these goals in mind, in closing we summarize what we have learned about self-declared climate change skeptics in the United States. We present overall implications of our work for climate change communication and policy and examine the role of a skeptic identity in the continued ideological fissures in US society and offer empirically founded ideas for change.

WHAT WE HAVE LEARNED ABOUT SKEPTICS

Our foray into understanding skeptics' beliefs has led us to the conclusion that climate change skepticism emerges as an opinion-based, stigmatized, social identity. It is formed and solidified in opposition to out-groups, which consists primarily of climate scientists but also members of the public who accept the veracity of climate science. Within this opinion-based identity exists several prominent intersecting categorical identities: political conservativism, conservative Christianity, whiteness, and masculinity.

Skeptics construct their self-concepts as social outsiders, unjustly marginalized and dismissed by society and ostracized throughout the life course. This marginalization happens simply because skeptics hold beliefs that do not align with majority views. Skeptics see themselves as reasonable individuals who simply pursue the truth about climate change. Yet this endeavor subjects them to constant "persecution" by those who hold opposing views about climate change (climate scientists and, to a lesser extent, all those who accept climate science).

We have identified several out-groups against whom skeptics construct their identity. Prominent among these out-groups is climate scientists. Given that climate scientists are the main group that generates and disseminates information about anthropogenic climate change, information that is largely incongruent with the beliefs of skeptics, skeptics may not only reject climate change information; they also develop an understanding of scientists as a clear, untrustworthy out-group. Skeptics believe that, contrary to evidence, scientists are unduly affected by funding and other incentives, they do not engage in objective science, they are resistant to the viewpoints of those who disagree with them, and they deliberately exclude those with opposing views.

It appears that identity-protective cognition is at play here: mechanisms through which individuals adopt beliefs that are shared by members of their salient in-group leads them to resist changes to these beliefs based on information stemming from perceived out-groups. For climate change skeptics whose skeptic identity has become salient, these processes serve to protect their status and self-esteem within the skeptic in-group. Skeptics evaluate information in ways that further strengthen their in-group belonging.[5]

While identities are important, they are inherently connected to underlying ideologies, the larger sets of ideas that create individual experiences of the world along with their attitudes and behavior. We examined two sets of ideologies, conspiracy ideation and religious ideation, as they impact the views and behaviors of climate change skeptics. We found that adhering to the existence of a conspiratorial "hoax" leads skeptics to reject the reality and human causes of climate change. Further, religious beliefs, such as human dominion or earth stewardship, shape their perceptions of climate change and responsibility toward the environment. Ideologies also shape skeptics' trust and emotional experiences in nuanced ways. For instance, skeptics with higher levels of religious and/or conspiracy ideation are less likely to trust science and scientists, and these ideologies serve to enhance or mitigate negative emotions associated with climate change and other environmental issues.

We also explore variance among climate change skeptics' environmental concerns and support for pro-environmental policies. We find that for some specific issues—air and water pollution, human waste, habitat destruction, and renewable energy—skeptics generally agree and support pro-environmental efforts. While they may not see issues like air and water pollution as connected to global warming, given the immediate threat of irreversible lethal climate change, perhaps the reasons need not matter as much as the shared interest in clean air and clean water. We further find that certain factors, such as experience with climate disasters, increase environmental concern and support for environmental issues among skeptics.

Additional examples of skeptics' identity-protective cognition and co-construction of out-groups are evident throughout our work. Similar mechanisms are at play in skeptics' distrust of the media, particularly mainstream and liberal media, which report on climate change information that is incongruent with their in-group beliefs. Skeptics view media as ideologically biased and unbalanced, driven by sensationalism and profit. Because skeptics exist largely within the conservative media echo chamber, they are exposed to messages that align with their preconceived beliefs. These processes, fueled by a powerful denial countermovement, ensures that US climate change skeptics are constantly exposed to information that protects their status and belonging within the skeptic in-group.

Skeptics evoke anger when thinking about climate change, and this anger is often directed toward perceived out-groups, such as scientists, liberal politicians, mainstream and liberal media, and environmental activists, all of whom they believe are engaged in a hoax to generate fear among the public or to accrue money and power. We demonstrate that skeptics with stronger (more salient) identities are more likely to feel anger in the face of climate change information reported by out-groups. This anger may further entrench skeptics in the conservative echo chamber, to allow them to avoid the discomfort of identity incongruent messaging. For skeptics with weaker convictions (doubters), worry and dread is commonplace when thinking about entities that they care about, so-called objects of care, such as pollution, habitat destruction, and animal species loss.

Taking into consideration these various identity dynamics, we proposed that climate change skepticism is best understood as a continuum. The continuum extends from *epistemic denial* at one end, where the strength and salience of the skeptic identity is at its highest, to *attribution doubt* at the opposite end, where the strength and salience of the skeptic identity is at its lowest. In congruence with this framework, we have demonstrated that the vast majority of skeptical beliefs fall somewhere along this continuum, including views about climate science, the levels of pro-environmentalism, conspiracy adherence, and the intensity of emotions expressed related to climate change.

Once adopted, the skeptic identity and its salience to self-concept largely determine skeptics' beliefs and behaviors toward climate change. Depending on the strength of skepticism, a skeptic may accept information that reaffirms a skeptic identity, may reject or dismiss information that is incongruent, may entrench further into the in-group, and may engage in behavior that reifies skeptic beliefs, such as seeking "alternative" sources of information on climate change.

COMMUNICATING WITH SKEPTICS

Based on our understanding of who climate change skeptics are, what drives their skeptic identity, and which ideological beliefs underly the skeptic identity, we propose a communication strategy that takes into consideration the continuum of skepticism and, where possible, delineate strategies that may work well with deniers from those that may motivate action among doubters.

Consider the Audience

When communicating the facts of climate change with skeptics it is important to consider the characteristics of the audience. We know from our work that skepticism operates as a continuum, from epistemic denial to attribution doubt. Given that epistemic deniers are more likely to be politically conservative (and embed themselves in a conservative media echo chamber), they have higher distrust of science, they are more prone to conspiracy and religious ideation, they are more likely to use anecdotes to reject climate change, and they are less likely to experience negative emotions related to climate change, it is plausible that the denier end of the spectrum will consist of skeptics who are least likely to be receptive to climate change communication. Deniers may show the strongest resistance to messages from climate scientists and others who are perceived as outgroups. For deniers, communication strategies may require a decoupling of objects of care from climate change, with messages focusing exclusively on things that they care about, such as pollution and deforestation.

Other skeptic groupings may respond to messaging of another sort. Epistemic doubters might resonate best with stories connecting human waste and trash with climate change and environmental destruction. Stories about renewable energy may be best packaged in the language skeptics use in their support of such initiatives, energy independence. Attribution doubters may respond better than others to stories about the actual science of climate change, given their relatively higher levels of trust in science. For religious members of the doubter group, stories about environmental stewardship and protecting God's creation may resonate best. Attribution doubters may respond best to stories about animal species loss, especially when accompanied by a message of agency and hope. Other messages targeting various categorical identities encompassing the points along the skepticism continuum (e.g., conservative political ideology) may also be effective. For example, emphasizing entrepreneurship over "big government solutions" for tackling climate change.

In short, there is no perfect story or frame that will reach all climate change skeptics. This population, despite holding a shared identity, is simply too varied in its engagement with climate science and media. Instead, making the messages personal may be important for both deniers and doubters. For example, regardless of their degree of skepticism, all skeptics appear to gravitate toward images of or stories about people who are seen as similar to themselves in age structure, race and ethnicity, values, and concerns.[6] Tailoring messages to the specific community or location may be useful, although this strategy may work better with doubters than deniers because of the tendency among deniers to disassociate local events and personal experiences from larger changes in the climate. Furthermore, given skeptics' higher tendency to discount the future with regard to climate change, focusing on things that may happen in the here and now might be more compelling than presenting information on long-term implications.

In the same vein, while all of us gravitate toward stories, doubters may be more receptive to narratives about people and places they care about that are affected by climate change or where climate solutions are working. Deniers may reject such stories or may even experience anger if such stories are perceived as deliberate attempts to manipulate their emotions. As a result, while "telling stories" is a preferred and effective communication strategy, attention should be paid to the specificities of the audience before employing it as a communication strategy.

Our pilot interviews with people who have changed their minds on climate change show one other possibility for successful climate communication: framing stories in a way that activates other identities held by skeptics (i.e., animal lover, college student, loving grandparent, good person, Christian steward, travel enthusiast). By making the counter-identity more salient, skeptics may decrease their skeptic-informed tendency to dismiss information about climate change and increase their willingness to listen and be open to certain possibilities.

Regarding climate change communication, a group that may be particularly hard to reach is outright deniers who believe in the existence of a conspiratorial hoax regarding climate change. For conspiracy adherents, debunking their beliefs by providing them with more facts (fact-based debunking) or by pointing out the flawed reasoning associated with conspiracy theories (logic-based debunking) may not be effective, as they are ideologically motivated to reject new information about climate change or challenges to their beliefs. Given that these skeptics exist, for the large part, within conservative echo chambers, such debunking efforts may even enhance their conspiratorial interactions with others existing in the same information systems. Psychologists Lewandowsky and Cook have recommended the use of trusted messengers to convey information to these skeptics.[7]

Consider the Messenger

Our interviews and surveys with self-declared skeptics show that, regardless of the strength of skepticism, most skeptics rely on media and information sources associated with the political right and harbor significant distrust of information sources associated with perceived outgroups. Because of the dominant political and religious ideologies that drive US climate change skepticism, skeptics are more likely to be receptive to messages arising from conservative political leaders, religious leaders, and other conservative elites who have garnered support and trust among skeptics.

One example of a bridge between skeptics who are ideologically conservative and the science of climate change is a relatively new organization called RepublicEN, spearheaded by former Republican South Carolina congressman Bob Inglis. Identifying as "ecoright," RepublicEN provides "a balance to the Environmental Left."[8] Declaring that "the age of conservative climate disputation is over," the organization purports to "believe in the power of American free enterprise and innovation to solve climate change." In other words, rather than continuing to debate the realities and human causes of climate change, the organization offers conservative solutions to tackle it.

RepublicEN and similar organizations operate under the theory that the conservative lack of engagement with climate change stems from the fact that traditional solutions to climate change go against conservative economic values. If such solutions are changed to better align with conservative values such as free-market ideology, then conservatives' support for climate action will increase.[9] This theory is supported by some of our survey data, where we observe high levels of support for investments in renewable energy and tax rebates for consumers who install renewable energy systems in their homes.[10]

Given that skeptics (mainly doubters) show some level of concern regarding future generations, young conservatives and Republicans may also be effective communicators of climate change among the political right. For example, the young conservative climate activist Benji Backer founded the American Conservation Coalition in 2017 and is exploring "market-focused responses" to the Green New Deal via the American Climate Contract.[11] Leading up to the 2020 US presidential election, Backer embarked on a forty-five-day road trip across nineteen states in an electric car. With this endeavor Backer's goal was to "highlight market-based environmental innovation and corporate stewardship as ways to fight the climate crisis in cities and towns across the country."[12]

Backer is not alone in this effort. Others, like Keira O'Brien, are advocating for carbon tax proposals to reduce greenhouse gas emissions via the Young Conservatives for Carbon Dividends initiative.[13] These young activists are active on Twitter and other social media platforms, spreading their message across conservative circles. Although the effectiveness of efforts to reach out to the political right remains unclear due to the endemic nature of climate change denial on the conservative side of the political spectrum, there is potential for such messages to reach deniers more readily than messages coming from perceived out-groups.

In this vein, Katharine Hayhoe's work is especially commendable. A prominent atmospheric climate scientist and the director of the Texas Tech University Climate Science Center, Hayhoe is an evangelical Christian. While evangelical Christians have significantly low levels of climate change acceptance, Hayhoe is in a unique position to be able to speak in language to which they might respond.[14] She attempts to depoliticize climate change through both her public speaking and her writing. For example, in her 2009 book, *A Climate for Change: Global Warming Facts for Faith-Based Decisions*, she and coauthor Andrew Farley argue that climate change is not about red or blue politics, but rather "about thermometer readings and history." An astute social media user, Hayhoe's 2018 TED talk, "The Most Important Thing You Can Do to Fight Climate Change: Talk about It," has accumulated close to four million views.

In short, these examples show the potential power of changing the meaning of a conservative identity to one that encompasses a level of care for the climate. We see the significance of identity salience, given the power of messages from perceived in-groups that make engagement with environmental issues more acceptable to those with shared identities.

Shape Emotional Engagement

Our work with skeptics has shown that emotions are an important consideration in climate change communication. Emotions are necessary for understanding the moral impacts of the risks of climate change as well as for motivating action. While emotions evoked by doubters (mainly worry and dread) and deniers (mainly anger) may be different in kind and intensity, moral emotions can play an important role in letting us assume responsibility for our own actions. As such, many emotion scholars recommend appealing to moral emotions to encourage the recipients toward "thorough ethical reflection about the impact of climate change."[15]

It is important to note here that evoking worry and other moral emotions may not work, and may even backfire, in the case of deniers. Deniers do not accept anthropogenic climate change and do not feel any ethical responsibility that would motivate them toward action. For deniers, communication techniques should decouple objects of care from climate change and instead focus attention on entities and events such as species loss, deforestation, and pollution.

For doubters, this strategy may work when worry is evoked to generate empathy toward those whom they care about, such as victims of extreme weather events or future generations. Our research shows that worry is associated with pro-environmentalism, so it is plausible that messages about objects of care that evoke worry or similar emotions may provide motivation to adapt behavior, especially when messages are offered with concrete narratives, images, and shared goals. We have argued elsewhere that "focusing on shared goals such as reducing pollution, preventing deforestation, and investing in renewable energy will benefit the climate and meet the concerns of skeptics."[16] Given climate change communications that generate alarm or fear can lead to information aversion, greater skepticism, and even denialism among skeptics, it is important to examine what specific objects trigger these emotions.[17] Ultimately, emotions should be considered as one element of a comprehensive communication strategy that targets skeptics, rather than an easy tool to be manipulated to generate desired behaviors.[18]

We wonder, too, about other emotions that have yet to be explored among climate change skeptics. In our interviews we witnessed the potential for nostalgia to be influential for people who want a better world for their children. Further, if skepticism may reflect a tendency toward information aversion in the face of frightening and complicated societal issues, to what extent might hope or agency influence skeptics' willingness to engage intellectually and emotionally with climate change?

There is no perfect recipe for the right emotions to tap in climate communication with all skeptics. Rather, the ideological and identity foundation held by skeptics shapes their engagement with information and action. Thus, particular emotional combinations may result in connection with some, but not all, skeptics.

Emphasize Common Ground and Agreed-Upon Solutions

For deniers who do not accept the physical realities, human causes, and impacts of climate change, and for doubters who are skeptical of these same things, dire predictions about uninhabitable climate futures may not motivate action. In fact, such fears of the future may even lead to information aversion and stronger levels of skepticism as a psychological coping strategy. Instead, communicators should employ strategies of some of the young conservative activists, who use a positive and hopeful focus on solutions and benefits upon which everyone can agree.

Skeptics earn disproportionate airtime in news stories and hold inordinate representation in political office. Given the difficulty in communication across political ideology and identity on partisan issues, perhaps the best strategy forward is not to try to convince them of the veracity of climate change, but rather to focus on areas of mutual support, regardless of the reasons that underlie concern.

For example, we find a high degree of concern about air pollution as well as a high level of support for clean energy (wind and solar).[19] Presenting clean energy as an environmental solution that both reduces air pollution (something that skeptics claim to care about deeply) and generates an economic opportunity (something that aligns with free-market values) may generate support among both deniers and doubters while simultaneously mitigating some of the effects of climate change. Our interview data suggests that emphasizing health benefits of climate solutions might still work with skeptics because they do care about air and water pollution, both of which have roots in fossil fuel-based production systems. Interview participant David succinctly stated it: "It's not like conservatives want to breathe dirty air!"[20]

Broadly, we agree with Wendy Ring in her recommendation to explain the difference between policy and individual action when communicating about climate change.[21] When people, especially skeptics, are asked to make changes to their behavior at the individual level or sacrifice/adjust their lifestyle choices, they may feel resistance, even anger. Particularly when engaging deniers, for whom climate change is a "fictitious problem," such requests for behavioral change may generate defensiveness. However, if climate solutions are presented as policies that generate benefits across the political spectrum (e.g., improved health and more economic opportunity), skeptics, doubters and deniers alike, may express more receptiveness to such messaging and solutions.

In short, we do not have time to wait for skeptics to get on board with climate change. The situation is dire and we need to capitalize on all and any political will to advance pro-environmental policies.

Minimize In-Group/Out-Group Tensions

While these suggested communication strategies may increase skeptics' receptiveness toward some climate solutions, they will not persuade self-declared skeptics about the science of climate change, especially when pertinent information is associated with perceived out-groups such as climate scientists, activists, and politicians. Skeptics who feel marginalized by those who accept climate science may simply refuse to engage in dialogue with the larger community.

Our findings suggest that skeptics feel as though they are deliberately excluded from scientific debate about climate change. One reason for this perception is their inability to access primary sources on the topic. To address this concern we advocate for more openly accessible science (scientific writing that uses clear, accessible language and is freely accessed without paywalls, including peer-reviewed publications). Regardless of whether the public at large has the technical skills, time, or resources to peruse available scholarship, such openness signals to the skeptic community that scientists are transparent, ethical, trustworthy, and willing to engage in public conservation. Furthermore, given the skeptic self-perception, skeptics may want to feel respected and treated fairly by the scientific community. Open science cultivates a sense of fairness and respect between the skeptical public and climatologists who analyze the issues at hand. Because climate scientists are the most likely threatening out-group regarding skeptics' identity-protective cognition, it is essential that they are provided with better training and professional resources to become effective science communicators who are able to communicate across broad swaths of the public. In this vein, we commend efforts such as the National Academies of Sciences science communication collection and the "science of science communication" colloquium series.[22]

Another strategy may be to increase the salience of the other identities held by skeptics that resonate more with climate change. When people feel attacked or marginalized for a particular identity, that identity becomes more resonant and important. Thus, instead of approaching conversation and action around climate change from an adversarial position, perhaps doing so from a place of shared ground or shared identity may best serve to bridge the identity divide. We note a series of environmental concerns that skeptics share with others, including the desire to preserve the national park system. Perhaps by first triggering more pro-environmental identities such as concerned parents, outdoorsmen, hunters, fans of Theodore Roosevelt, or Christians, or connecting at points of agreement, communication between skeptics and those who accept climate science could be more effective.

ADVANCE POLICIES WITH CROSS-PARTISAN SUPPORT

The ideas that we have presented for communicating climate change with skeptics can address the "attitude roots" of skepticism (e.g., their underlying ideologies, identity needs, and emotions). Extending their work on "attitude roots," psychologists Matthew Hornsey and Kelly Fielding recommends "Jiu Jitsu persuasion" as a way to communicate science with skeptics. The goal of this persuasion technique is to "identify the underlying motivation, and then tailor the message so that it aligns with that motivation."[23] Our suggested communication strategies, such as tapping into the free-market ideology of conservative-leaning skeptics or presenting messages by in-group sources, largely aligns with this technique. This framework also suggests that scientists like Hayhoe would be more effective at communicating climate change if they emphasize a shared identity (religious identity, in her case). The same argument would suggest that messages that focus on attitudes and behavior seen as normative for a group may garner more support: "People like you care about pollution and want to regulate it."

Skeptics care about a number of environmental issues. Prominent among these are pollution, habitat destruction, and species loss. In fact, skeptics go as far as expressing a desire for federal regulation for accomplishing their environmental goals, even though regulatory policies run counter to many of their conservative and libertarian political positions. Given that skeptics' environmental goals reflect a desire to protect and conserve specific objects of care that elicit worry and concern (especially among doubters, and to a lesser degree among deniers), policies that target pollution reduction or species/habitat conservation might garner wider support among skeptics. Needless to say, associated policies for the regulation of atmospheric carbon dioxide or species conservation will also tackle climate change by reducing overall greenhouse gas emissions.

We also note significant support among skeptics for investments in solar and wind power. Fewer than half of the skeptics we surveyed support increasing use of fossil fuels via increased fracking or extending offshore drilling for oil. Energy independence is a topic that garners wide support among conservatives in general. Given that a significant fraction of skeptics in the US lean conservative, highlighting climate solutions that are coupled with renewable and independent energy futures could bring more skeptics on board, toward enacting energy-related climate policies.

Furthermore, we observe that the psychological distance between skeptics and climate change need not be bridged for skeptics to be able to engage in pro-environmental behavior and support pro-environmental policies. As a result, policymakers should be able to move forward quickly on policies that address issues that skeptics do care about, which will lead to protecting the environment in the short term and mitigating climate change in the long-term.

However, when discussing these climate and energy solutions, it is important that skeptics' concerns are adequately acknowledged and addressed. Skeptics raise legitimate concerns about alternative energy sources, such as the cost of solar panels and electric vehicles. Ultimately, to bring skeptics along to enacting climate policy, it is essential that they feel their viewpoints are being taken seriously and addressed sufficiently.

It is also critical that those who implement policy, including Republican political leaders, are made aware of the extent of support among skeptics for pro-environmental and pro-climate policies. If Congress operates with certain misconceptions about the level of public support and public demand for energy policy or environmental protection, the pace at which leaders move forward and the policies they enact will continue to fail to adequately address skeptics' and nonskeptics' demands for a cleaner, safer environment.[24]

"EXITERS": CHANGING THE SKEPTIC IDENTITY

While preliminary, our work uses interviews with former skeptics to offer important insights on what appears to change skeptics' minds about climate change. We identified two trajectories through which skeptical perspectives may change: "denial to acceptance" and "doubt to acceptance." We find that denial to acceptance occurs due to a larger ideological shift caused by changing personal circumstance and formation of a new identity—for example, changing one's geographic location, forming a new friendship, attending college, or coming out—circumstances that are not easy to manipulate to gain a desired effect on one's beliefs. We simply cannot control an individual skeptic's geographic or social location or provide them with specific personal examples. However, it is possible to frame conversations in ways that minimize the salience of the skeptic identity and increase that of other identities, which may create new in-group unities across ideological divides and open up skeptics to new information on climate change. Adjusting the salience of an identity can provide transformative power for overcoming deeply entrenched racist and religious bias. Research by political scientist Ala' Alrababa'h and his team find that simply having exposure to positive role models from an out-group may increase acceptance across ideologies.[25]

Consider the second group, those who went from doubt to acceptance. The two most significant factors that changed their beliefs about climate change were learning about climate change from a respected authoritative figure or having direct experience with its effects. We describe these required cascades as a "magical aligning of events" that would be difficult, if not impossible, to replicate across large groups. However, the lessons we have learned from interviewing people who admit to having changed their minds about climate change does provide some insights on what may shake the salience of the skeptic identity.

Most people across the world are no longer insulated from the impacts of climate change. The effects of a warming planet are being experienced by the poorest residents of coastal Bangladesh and the wealthiest residents of Malibu, California, exposing unique geographical and social vulnerabilities. As extreme weather events such as storm surges, cyclones, and wildfires become more commonplace, more and more people, including skeptics, will encounter direct personal evidence of or hear more frequently about friends and family who are affected by the effects of climate change. At a certain point it may become untenable to reject the evidence in front of one's own eyes. However, we simply cannot wait for everyone to experience the worst effects of climate change before we act.

Given the difficulty of changing salient identities, one strategy that will lead to greater political will on climate change is to focus on the 14 percent of Americans who remain unsure of the phenomenon. Attention should focus on delivering accurate climate change information to this group via trusted information sources: not just news media and scientists, but also elders, community leaders, college professors, and those who are savvy in newer media such as social media and the blogosphere, where skeptics tend to coalesce. Our understanding of climate change skepticism as a continuum suggests that well-tested communication strategies— properly framing the message, personalizing the message, evoking emotions, and so forth—must target audiences based on their degree of skepticism.

For instance, we may target environmental/climate campaigns to college students, a social location within which skeptics claim to having changed their minds on the topic. We should consider enlisting religious leaders, as respected community members whose ideologies align with those of skeptics, to deliver messages on the urgency of the issue. The energy and resources spent on debating climate change facts with deniers may be better utilized by redirecting it toward doubters who are willing to engage with the topic. Climate campaigns can harness the power of personal environmental experiences and emotions to motivate action by using newer techniques like virtual reality, documentaries, and narrative fiction, which together could enhance engagement with the topic at hand as well as increase environmental concern and policy support among doubters. These media sources are known to increase empathy and prosocial behavior among those who experience them, even and especially toward out-groups.[26] In the rare cases where we do encounter former skeptics who went from denying to accepting climate change, their voices may offer a powerful force to climate change communication campaigns because they are uniquely positioned to speak to the very concerns of staunch skeptics. This idea comes from research into extremism and deradicalization, which suggests that counter messages from former members of extremist groups ("exiters") may be more effective at changing attitudes than other information sources.[27]

In short, we see the fundamental need of shifting the climate change skeptic identity to facilitate a shift in the political will. Among deniers, this identity is more salient and will require greater effort to produce a shift. Among doubters, who may have less salient skeptic identities, it may be possible to cultivate empathy and knowledge through media systems such as virtual reality and narrative fiction, two sources known to produce empathy that may also mimic the effects of firsthand experience with climate disasters. In either case, challenging the skeptic identity by increasing the salience of a second identity more amenable to prosocial and pro-environmental thought may open people to new ways of looking at climate change.

We may not be able to fully beat denialism, but changing minds, even gradually, may lead to changing policies if enough people join the effort via personal behavioral changes, direct climate actions, or voting decisions.

CONCLUSION

Without swift action on climate change, scientists predict irreversible and lethal consequences in the near future. Yet we lack the political and public will for politicians to take necessary action. Skeptics are overrepresented in government and have a strong hold on US policy development. Moreover, as the capitalist machine grinds on, national leaders continue to focus on economic growth at the peril of humanity's survival.

We trust in the ability of Americans to build collaborative concern about climate change across the skeptic/nonskeptic identity divide. We do not purport to think that such cross-ideological communication will be easy or swift but we do believe it is possible. To start, political leaders need to advance pro-environmental and pro-climate policies upon which people agree: regulations to address pollution, preserve habitat, and support investment in renewable energy systems. We do not need to, nor should we wait for, consensus on climate change to take these actions.

Concurrently, we believe tailored strategies for communication across the identity divides can be effective. We particularly see that the construction and magnification of counter-identity systems is key to increasing empathy, connection, and opening people's minds to considering climate science and climate action.

Jay Van Bavel contends that "truly great leaders are capable of rallying people around a common identity." Under the Trump administration, identity lines dividing the American public, especially among those aligned with political ideologies, deepened and became entrenched. Current leaders have a challenge to diffuse this in-group/out-group division and unite the public around a shared sense of "Americanness." In the 1960s John F. Kennedy rallied Americans around the goal of reaching the moon. He used the power of scientific knowledge to bring citizens together and "get people around a common, visionary purpose." We should use the current moment and the dire implications facing humanity to "motivate people to feel a common sense of purpose, to make sacrifices and help one another, and move away from trolling one another online or in other forms of media."[28] Addressing the effects of climate change must be our next shared purpose.

appendix

METHODOLOGY

OUR PROJECT BEGAN IN EARNEST the summer of 2017 when we secured a small seed grant from the University of Idaho to conduct interviews with climate change skeptics. Working with a team of University of Idaho student researchers—Ashli George, Sarah Olsen, Kathryn Pawelko, Christine Sedgwick, and Hannah Spear—we interviewed thirty-three people in Idaho who self-identified as skeptical about climate change. These interviews, which lasted between 30 and 120 minutes and were held in public coffee shops, fell into the category of what social scientists call "semi-structured," meaning we had a list of questions we wanted to ask but tried our best to let the conversation flow naturally over the various topics. To find people to interview we posted fliers in grocery stores throughout the state of Idaho and placed ads in commercial Facebook groups (e.g., Buy, Sell, Trade of [town name] community groups) throughout the state.

We began each interview by asking the participants their broader perspectives on climate change, with several probing questions to get at the heart of their beliefs on this subject. We also asked about a variety of environmental topics—pollution, the ozone hole, melting ice caps, extreme weather patterns—and the emotions they feel when they think about climate change. Our interviews included additional questions about participants' relationship to science, to scientists, to media, and to various information sources. We also asked about the subjects' personal experiences growing up in their respective communities.

In total we interviewed thirty-three skeptics in this first round, thirty-two of whom reside in Idaho and one who lives just across the border in

Washington. Seventeen participants were from the southern half of Idaho and the other sixteen were from the north half of the state. Throughout this book, we use pseudonyms when referring to the participants to protect their privacy. More details on the specific demographic backgrounds of our participants such as age, gender, race, political and religious affiliations, and education are available in table 1.

The interview process ended once we reached "saturation," the point at which the information provided was no longer new and the researchers were certain they had collected all the data needed to understand the phenomenon under study.[1] After completing the interviews we transcribed them verbatim and developed what social scientists refer to as "codes," or a system of identifying patterns and themes that manifest from the data when using inductive analysis.

It is important to note that in the use of grounded theory, analysis is nonlinear. Researchers collect data, develop codes, analyze data, and compare their work to existing scholarship throughout the process. New ideas and findings lead researchers to search for and collect additional data to enhance their understanding. This process is repeated until researchers feel they have reached the point of saturation.[2]

To code our data, each researcher read the interviews searching for evident patterns. From these patterns we developed a series of theoretical, indicative, and semi-indicative codes. Using grounded theory, initial coding tends to be indicative; it identifies patterns and phenomena. During this stage the coding consisted of literal interpretations of participants' words—for example, that climate change is a hoax or that Al Gore is making up climate change to make money. As we sifted through the interviews, we talked with each other and found connections between ideas that led to more sophisticated theoretical understandings of skepticism—such as climate change skepticism as an identity.

We applied this coding system to the data set to identify patterns and to begin the process of theory-building. To test the reliability of our coding system, we began by independently coding one randomly selected interview and assessing the effectiveness and accuracy of the codes, then adjusted our operationalization of the codes as needed. We then randomly selected five additional interviews, independently coded them, and tested for intercoder reliability using the qualitative data analysis and code comparisons functions of NVivo software. We observed over 95 percent agreement on nearly all our resulting codes and fewer than 5 percent of the codes reported disagreement. Once this intercoder reliability was established, the authors independently coded the remaining interviews.

Once we completed the analysis of our interview data, we pored through our findings and existing scholarship to develop an empirically driven set of survey questions that could assess the perspectives of a broad swath of climate change skeptics. Our research team for this step consisted of an environmental humanities scholar, Dr. Jennifer Ladino (University of Idaho); two geographers, Dr. Steven Radil (US Air Force Academy) and Dr. Thomas Ptak (Texas State University); University of Idaho student researcher Randolph P. Stuart; and ourselves.

We administered the online survey to adults living in Washington, Oregon, and Idaho. For this process we hired Qualtrics, a company that specializes in conducting representative online surveys. Quota constraints for gender and education were applied to our sample to be representative of US census data for the Pacific Northwest region. The survey was first tested and adjusted in September 2019. The full data collection phase took place between November 2019 and January 2020. The final sample (*N*=1000) consists of only respondents who met the identified constraints, completed all survey questions, and passed the speeding and validity checks.

The survey begins with a consent validation question followed by two screening questions. The screening questions asks a respondent to report whether they believe (1) "climate change is happening" and (2) "climate change is caused by human activities" (response categories: Yes=1; No=2; Not sure=3). Respondents who said that they believed climate change is happening and it is caused by human activities (i.e., respondents who answered yes to both screening questions) were screened out, thus limiting the sample to only those who expressed denial or uncertainty regarding anthropogenic climate change. Participants who met the screening criteria were then asked to answer the full survey, which consists of questions on their beliefs regarding climate change, the environment, science and scientists, information sources, energy production and use, policy preferences, their emotions, and demographic factors. Information on the demographic characteristics of the full survey sample can be found in table 2.

Concurrent with our survey collection, we undertook a second round of interviews with people who had changed their minds about climate change. Our research team for this step of the process consisted of Randolph P. Stuart, University of Idaho sociology alumnus Amber Ziegler, Jennifer Ladino, Thomas Ptak, Steven Radil, and us. This time, we interviewed twenty-one people who self-identified as having changed their minds about climate change. With this work we sought to uncover the factors that contribute to changing one's mind about controversial and heavily politicized issues such as climate change.

To find participants for this round of interviews we again posted notices in online Facebook Marketplace sites. We also shared invitation emails across the state of Idaho via various university extension offices. Perhaps unsurprisingly, it proved quite difficult to find people who had changed their minds about climate change. As a result, this data collection process lasted for almost two years, from the end of 2018 through the start of 2020.

The twenty-one identified participants all hailed from Idaho or nearby in Washington. While some participants began as deniers (outright rejecting climate change) but became believers in climate change (n=6), others experienced more subtle changes in their thinking or went from believing climate science to rejecting it. Six participants were doubters (uncertain about climate change) who became believers. (Using our umbrella concept of "skeptic," this indicates that twelve skeptics moved from doubt or denial to believing in climate change.) One participant went from fully denying climate science to becoming a doubter. Two moved from being strong believers in climate change to having an increase in doubt. Four moved from being moderate believers to strongly accepting climate science. One went from being a doubter to becoming a denier, while another one shifted in their emotional experience with climate change—moving from feeling hopeful about solving climate change to feeling increasingly defeated about it. Additional demographic characteristics of these participants can be found in table 1.

In the second round of interviews we began by asking participants to tell us about themselves: where they grew up and what their communities and families were like. We then asked them about their early beliefs about climate change: where, when, and how they developed them; how often they thought about climate change; and the role religion plays in their perspective. Next we asked participants to talk about the period when they changed their minds about climate change—the people and places instrumental in this shift and why it occurred. This was followed by a discussion about the nature and nuances of their current beliefs. We ended by asking participants about their personal engagement with the issue of climate change—its impact on their lives or their activism related to it; their support for environmental policies; and the emotions they experience when discussing climate change. Once the interviews were completed, we replicated the transcription, coding, and analysis used during the first phase of interviews.

TABLE 1. *Pseudonyms and demographic information of interview participants*

ROUND 1 PARTICIPANTS

Pseudonym	*Age*	*Gender*	*Race/ Ethnicity*	*Political affiliation*	*Religious affiliation*	*Education*
Allen	22	Man	White	unaffiliated	mainline Christian	some college
Ben	20s	Man	White	moderate conservative	nonreligious	some college
Bill	40s	Man	White	Libertarian	nondenominational Christian	college degree
Blake	mid- 30s	Man	White	unaffiliated	NS	some college
Bob	20s	Man	White	anarchist	NS	some college
Brent	28	Man	White	conservative	LDS	college and advanced degree
David	60s	Man	White	conservative	Evangelical	college and advanced degree
Douglas	40s	Man	White	apolitical	Christian	college degree
Greg	47	Man	White	NS	NS	NS
Henry	34	Man	White	conservative Republican	NS	college degree

Jack	50s	Man	White	Democrat	NS	high school
Jake	40s	Man	White	Libertarian, grew up Republican	not religious, grew up Fundamentalist Christian	college degree
James	64	Man	White	Libertarian	Evangelical	college degree
Jane	mid-60s	Woman	White	conservative	NS	NS
Jennifer	33	Woman	White	Independent	NS	high school
Jill	20s	Woman	White	Republican	Catholic	some college
Jodie	20	Woman	Latino	Republican	Catholic	some college
Joe	NS	Man	White	unaffiliated	NS	NS
Karen	35	Woman	White	Republican	LDS	college degree
Lee	20	Man	White	Republican	Evangelical	some college
Logan	late 50s	Man	White	moderate conservative	mainline Christian	college degree
Mark	63	Man	White	Libertarian	atheist	college degree
Matt	NS	Man	White	Republican/ Independent	NS	college degree
Nancy	60s	Woman	White	apolitical	NS	college degree
Nick	60	Man	White	conservative Democrat	NS	associate degree
Pam	over 60	Woman	White	Independent	NS	college degree
Peter	NS	Man	White	leans Republican	NS	college degree
Ron	50s	Man	White	"hesitant" Republican	NS	NS
Sabrina	20s	Woman	White	Republican	LDS	college degree
Sam	30	Man	White	skeptical of both parties	Catholic	NS

Savanah	20s	Woman	White	"on the line"	LDS	some college
Tyler	late 30s	Man	White	NS	NS	NS
Zed	50s	Man	White	Republican	NS	NS

ROUND 2 PARTICIPANTS

Pseudonym	*Age*	*Gender*	*Race*	*Political affiliation*	*Religious affiliation*	*Education*
Anna	early 30s	Woman	White	changed views, conservative to liberal	Evangelical until college	master's degree
Alex	25	Nonbinary	White	changed views, conservative to liberal	Evangelical until just after college	master's degree
Brent	77	Man	White	NS	Lutheran until college, now agnostic	bachelor's degree
Colton	mid-30s	Man	White	changed views, conservative to liberal	left church during graduate school	master's degree
Eli	mid-20s	Man	White	NS	Lutheran	in graduate school
Greg	mid-40s	Man	Native American	NS	non-Christian, traditional beliefs of affiliated tribe	bachelor's degree
Henry	32	Man	White	moderate liberal	nonreligious	bachelor's degree
James	28	Man	White/ Latino	changed views, conservative to liberal	Evangelical until college	bachelor's degree
Max	19	Man	White	changed views, conservative to liberal	Evangelical until college	in college
Samantha	mid- to late 20s	Woman	White	anarchist	atheist	bachelor's degree

NOTE NS (not stated) are those who did not specify characteristics during their interviews. In some cases age, gender, and race are inferences made by the interviewer. When mentioned by the participant, self-disclosed categories are used.

TABLE 2. *Survey sample and sub-samples across skepticism continuum (% or mean)*

SAMPLE CHARACTERISTIC	FULL SAMPLE	EPISTEMIC DENIER (18%)	EPISTEMIC DOUBTER (22.6%)	ATTRIBUTION DENIER (18%)	ATTRIBUTION DOUBTER (34.2%)
AGE (%)					
18–29	20.7	14.5	26.9	19.4	21.4
30–39	14.3	13.4	15.0	10.2	15.2
40–49	10.0	10.1	12.4	8.0	8.5
50–59	15.0	14.5	13.7	17.0	14.9
> 60	40.0	47.5	31.9	45.4	40.0
GENDER (%)					
Man	49.0	62.0	40.3	58.0	44.2
Woman	49.9	35.8	58.4	41.5	55.3
Other	1.1	2.2	1.3	0.6	0.6
EDUCATION (%)					
No high school diploma or equivalent	11.7	7.8	20.4	8.5	10.5
High school diploma, GED, or equivalent	19.4	19.6	25.2	15.9	16.7
Some college	21.5	21.8	16.8	19.3	24.6
Associate degree	8.5	12.3	9.3	7.4	6.7
Bachelor's degree or higher	38.9	38.5	28.3	48.9	39.3
ANNUAL INCOME (%)					
< $25,000	24.9	21.2	30.5	18.8	26.3
$25,000–$49,000	28.3	31.3	28.8	26.1	29.2
$50,000–$75,000	17.8	20.7	14.6	18.2	16.4
> $75,000	29.0	26.8	26.1	36.9	28.0

SAMPLE CHARACTERISTIC	FULL SAMPLE	EPISTEMIC DENIER (18%)	EPISTEMIC DOUBTER (22.6%)	ATTRIBUTION DENIER (18%)	ATTRIBUTION DOUBTER (34.2%)
RELIGIOSITY (MEAN) (NEVER ATTENDS [=1] TO ATTENDS MORE THAN ONCE A WEEK [=7])					
Religiosity	3.0	3.7	2.75	3.09	2.84
POLITICAL IDEOLOGY (%)					
Liberal	15.6	7.9	15.9	12.4	22.5
Moderate	42.6	29.6	51.8	35.8	48.0
Conservative	41.8	62.6	32.3	51.7	29.5
RACE (%)					
White	88.8	88.8	88.5	88.1	88.9
Other	11.2	11.2	11.5	11.9	11.1
CONSPIRACY IDEATION ("CLIMATE CHANGE IS A HOAX") (%)					
Yes	24.5	76.0	7.1	29.5	2.0
No/Not sure	75.5	24.0	92.9	70.5	98.0
TRUST IN MAINSTREAM MEDIA (MEAN) (HARDLY AT ALL [=1] TO A GREAT DEAL [=3])					
Trust in mainstream media	1.33	1.23	1.27	1.36	1.43
CLIMATE CHANGE INFORMATION RELIANCE (MEAN) (NOT AT ALL RELY [=1] TO RELY A GREAT DEAL [=7])					
Fox Network/Fox News	2.79	3.31	2.49	3.01	2.55
Infowars	1.55	2.05	2.16	2.14	1.94
MSNBC	1.60	1.93	2.25	2.08	2.30
CNN	1.66	1.92	2.30	2.09	2.43
NPR	1.74	2.04	2.41	2.26	2.72
One America News	1.64	2.13	2.33	2.25	2.08

NOTE All mean/proportion comparisons across skeptic types are significant at $p < .01$ except income, race, and three of the information sources (Infowars, MSNBC, One America News).

TABLE 3. *Selected beliefs across the skepticism continuum*

	EPISTEMIC DENIERS	EPISTEMIC DOUBTERS	ATTRIBUTION DENIERS	ATTRIBUTION DOUBTERS	ANOVA SIGNIFICANCE
CLIMATE CHANGE BELIEFS (STRONGLY DISAGREE [=1] TO STRONGLY AGREE [=7])					
What some people call climate change is really just natural variation	5.61 (1.70)	4.45 (1.09)	5.49 (1.50)	4.28 (1.33)	F=35.58 p < .001
There is no way to determine whether human activities cause climate change	4.72 (1.83)	3.91 (1.09)	4.25 (1.66)	3.57 (1.29)	F=19.03 p < .001
My personal experience tells me that climate change is not real	5.24 (1.74)	3.81 (1.03)	3.20 (1.58)	2.71 (1.27)	F=87.31 p < .001
Potential negative effects of climate change have been exaggerated	5.63 (1.73)	4.33 (1.16)	4.91 (1.75)	3.79 (1.47)	F=43.64 p < .001
There is no scientific consensus that human-caused climate change is happening	5.31 (1.55)	4.02 (1.02)	4.73 (1.62)	3.61 (1.29)	F=46.56 p < .001
SCIENCE ATTITUDES (STRONGLY DISAGREE [=1] TO STRONGLY AGREE [=7])					
Climate scientists do not have enough data to know that human-caused climate change is happening	5.27 (1.60)	4.22 (1.10)	4.79 (1.56)	3.82 (1.36)	F=34.91 p < .001
Climate modeling isn't science	4.88 (1.43)	3.84 (1.03)	3.93 (1.59)	3.34 (1.29)	F=33.89 p < .001
Climate scientists ignore those who disagree with them	5.34 (1.57)	4.30 (1.21)	4.78 (1.59)	4.06 (1.37)	F=22.61 p < .001
Scientific journals only publish papers that conclude climate change is happening	4.98 (1.50)	4.15 (1.15)	4.44 (1.51)	3.78 (1.43)	F=19.77 p < .001
Climate scientists are influenced by funding	5.59 (1.46)	4.53 (1.29)	5.19 (1.55)	4.38 (1.41)	F=22.49 p < .001

	EPISTEMIC DENIERS	EPISTEMIC DOUBTERS	ATTRIBUTION DENIERS	ATTRIBUTION DOUBTERS	ANOVA SIGNIFICANCE
ENVIRONMENTAL CONCERNS (NOT AT ALL CONCERNED [=1] TO VERY CONCERNED [=7])					
Air pollution	3.98 (1.91)	4.62 (1.62)	4.82 (1.59)	5.59 (1.36)	F=27.72 $p < .001$
Declining water quality	4.03 (1.98)	4.67 (1.82)	4.84 (1.67)	5.52 (1.48)	F=20.30 $p < .001$
Habitat destruction	3.87 (1.88)	4.49 (1.66)	4.88 (1.69)	5.39 (1.48)	F=22.16 $p < .001$
Animal species extinction	3.75 (1.96)	4.54 (1.69)	4.57 (1.70)	5.36 (1.56)	F=22.33 $p < .001$
Sea levels rising	2.77 (1.76)	3.97 (1.60)	4.03 (1.79)	4.93 (1.56)	F=42.88 $p < .001$
ENVIRONMENTAL POLICY SUPPORT (DOES NOT SUPPORT [=1] TO SUPPORTS A GREAT DEAL [=7])					
Investment in solar panel farms	3.37 (2.38)	3.68 (2.43)	4.33 (2.18)	4.79 (2.19)	F=12.19 $p < .001$
Investment in wind turbine farms	3.14 (2.26)	3.86 (2.37)	4.22 (2.30)	4.92 (2.12)	F=17.09 $p < .001$
Government regulation of fuel efficiency standards in new cars	3.06 (2.17)	3.48 (2.41)	3.95 (2.17)	4.45 (2.20)	F=11.63 $p < .001$
Federal regulations over water pollution	3.91 (2.39)	3.94 (2.54)	4.76 (2.02)	5.15 (2.08)	F=11.62 $p < .001$
Federal regulations over air pollution	3.63 (2.24)	3.73 (2.44)	4.39 (2.05)	4.98 (1.99)	F=13.92 $p < .001$
Withdrawal from Paris Climate treaty	4.34 (2.76)	2.56 (2.33)	3.93 (2.78)	2.66 (2.44)	F=16.61 $p < .001$
Cutting funding for EPA	3.68 (2.35)	2.28 (2.09)	3.70 (2.35)	2.65 (2.03)	F=13.97 $p < .001$

TABLE 3.
Continued

	EPISTEMIC DENIERS	EPISTEMIC DOUBTERS	ATTRIBUTION DENIERS	ATTRIBUTION DOUBTERS	ANOVA SIGNIFICANCE
EMOTIONAL RESPONSES (DOES NOT AT ALL FEEL [=1] TO FEELS AN EXTREME AMOUNT [=7])					
Worry	1.99 (1.51)	2.75 (1.42)	2.57 (1.56)	3.38 (1.69)	$F=20.51$ $p < .001$
Dread	1.91 (1.51)	2.42 (1.45)	2.24 (1.53)	2.91 (1.69)	$F=10.80$ $p < .001$
Sadness	2.08 (1.64)	2.72 (1.54)	2.55 (1.64)	3.26 (1.71)	$F=13.79$ $p < .001$
Grief	1.86 (1.48)	2.32 (1.49)	2.13 (1.50)	2.66 (1.69)	$F=7.38$ $p < .001$

NOTE In the first four columns, the top number is the mean and the number in parentheses is the standard deviation.

notes

INTRODUCTION

1 Yale Climate Opinion Maps, 2021.
2 McCright and Dunlap, "Anti-Reflexivity."
3 McCright and Dunlap, "Anti-Reflexivity."
4 Lahsen, "Experiences of Modernity in the Greenhouse"; Dunlap and McCright, "Organized Climate Change Denial"; Dunlap and McCright, "Challenging Climate Change"; Dunlap and Jacques, "Climate Change Denial Books."
5 Macinnis and Krosnick, "Climate Insights 2020."
6 McCright and Dunlap, "The Politicization of Climate Change"; McCright and Dunlap, "Cool Dudes."
7 McCright, "Anti-Reflexivity and Climate Change Skepticism."
8 Hamilton, "Did the Arctic Ice Recover?"; Hamilton, Cutler, and Schaefer, "Public Knowledge and Concern."
9 Leiser and Wagner-Egger, "Determinants of Belief."
10 Rahmstorf, "The Climate Sceptics," 77.
11 Engels et al., "Public Climate-Change Skepticism"; McCright and Dunlap, "The Politicization of Climate Change"; McCright and Dunlap, "Cool Dudes"; McCright, "Anti-Reflexivity and Climate Change Skepticism," 78.
12 Haltinner and Sarathchandra, "Considering Attitudinal Uncertainty."
13 NASA, "Global Warming vs. Climate Change," 2022.
14 Nisbet, "Communicating Climate Change," 15.
15 Haltinner and Sarathchandra, "Pro-Environmental Views of Climate Skeptics."
16 For more details about our data and analytical procedures, see the methodology section in the appendix.
17 These data analysis procedures are described in greater detail in the appendix.
18 Yale Climate Opinion Maps, 2021.
19 Boaz, *Libertarianism.*
20 CSPN, "Lesson One: Who Belongs."
21 Jones, "Conservatives Greatly Outnumber Liberals."

22 Lipka and Wormald, "How Religious Is Your State?"
23 CSPN, "Lesson Fourteen: Industrialization, Technology, and Environment in Washington"; Pilgeram, *Pushed Out.*
24 Lee, *The Making of Asian America.*
25 Dietrich, *The Final Forest*; Seideman, *Showdown at Opal Creek*; *The Oregonian*, "Oregon Standoff Timeline."
26 Pilgeram, *Pushed Out.*
27 Shapiro and Karlinsky, "Militia, Along with Family of Cliven Bundy."
28 Glaser and Strauss, *The Discovery of Grounded Theory*; Luker, *Salsa Dancing in the Social Sciences*; Ragan, *The Comparative Method*; Haltinner, *No Perfect Birth.*
29 Allport, "The Historical Background of Social Psychology."
30 Stets and Serpe, "Identity Theory."
31 Hogg and Abrams, *Social Identifications*; Stets and Burke, "Identity Theory and Social Identity Theory."
32 Burke and Tully, "The Measurement of Role/Identity"; Thoits, "Multiple Identities."
33 Stets and Burke, "Identity Theory and Social Identity Theory."
34 Turner et al., *Rediscovering the Social Group.*
35 Stets and Burke, "Identity Theory and Social Identity Theory."
36 Hogg, Terry, and White, "A Tale of Two Theories"; Stets and Burke, "Identity Theory and Social Identity Theory.".
37 Turner and Giles, *Intergroup Behavior*; Turner et al., *Rediscovering the Social Group*; Hogg and Abrams, *Social Identifications.*

ONE *Skepticism as a Stigmatized Identity*

1 Tajfel, *Differentiation between Social Groups*; Tajfel and Turner, "Social Identity Theory of Intergroup Behavior."
2 Stets and Burke, "Identity Theory and Social Identity Theory"; Burke and Stets, *Identity Theory*; Tajfel and Turner, "Social Identity Theory of Intergroup Behavior."
3 Tajfel and Turner, "Social Identity Theory of Intergroup Behavior."
4 Stets and Burke, "Identity Theory and Social Identity Theory."
5 Hackel et al., "From Groups to Grits"; Vedantum, "Group Think."
6 McGarty et al., "Collective Action."
7 Bliuc et al., "Public Division about Climate Change."
8 Brewer, "The Psychology of Prejudice"; Branscombe et al., "The Context and Content of Social Identity Threat."
9 Goffman, *Stigma.*
10 Goffman, *Stigma*; Quinn, "Concealable versus Conspicuous Stigmatized Identities."

11 Stafford and Scott, "Stigma, Deviance and Social Control."
12 Goffman, *Stigma*; Stafford and Scott, "Stigma, Deviance and Social Control."
13 Yale Climate Opinion Maps, 2021.
14 Hornsey and Fielding, "Attitude Roots and Jiu Jitsu Persuasion," 466.
15 Hornsey and Fielding, "Attitude Roots and Jiu Jitsu Persuasion," 467; Branscombe, Schmitt, and Harvey, "Perceiving Pervasive Discrimination," 135.
16 Sarathchandra, Haltinner, and Grindal, "Climate Skeptics' Identity Construction."
17 Brodsky, "Constructing Deniers."
18 Note that these interviews occurred in 2017 during the Trump administration.
19 Rahmstorf, "The Climate Sceptics."
20 Hornsey and Fielding, "Attitude Roots and Jiu Jitsu Persuasion," 467.
21 Harris and Brouwer, "Climate Scientists Claim."
22 Harris and Brouwer, "Climate Scientists Claim."
23 Harris and Brouwer, "Climate Scientists Claim."
24 Hornsey and Fielding, "Attitude Roots and Jiu Jitsu Persuasion," 467.
25 Rahmstorf, "The Climate Sceptics."
26 Del Vicario et al., "Modeling Confirmation Bias and Polarization."
27 Stets and Burke, "A Sociological Approach to Self and Identity."
28 Stets and Burke, "A Sociological Approach to Self and Identity."
29 Stets and Burke, "A Sociological Approach to Self and Identity."
30 Iyengar and Hahn, "Red Media, Blue Media."
31 Iyengar, Sood, and Lelkes. "Affect, Not Ideology."

TWO *Dis(Trust) of Science*

1 Leiserowitz et al., "Climategate, Public Opinion."
2 NSF, "The State of U.S. Science and Engineering 2020."
3 Jones, "Democratic, Republican Confidence."
4 NSF, "The State of U.S. Science and Engineering 2020."
5 Bugden, "Denial and Distrust."
6 NSF, "The State of U.S. Science and Engineering 2020."
7 Sarathchandra and Haltinner, "A Survey Instrument," 18.
8 American Academy of Arts and Sciences, "Perceptions of Science in America."
9 Yale Climate Opinion Maps, 2021.
10 Leiserowitz et al., "Global Warming's Six Americas."
11 Brewer and Ley, "Whose Science Do You Believe?"
12 Almassi, "Climate Change, Epistemic Trust."
13 Malka, Krosnick, and Langer, "The Association of Knowledge."
14 Malka, Krosnick, and Langer, "The Association of Knowledge."

15 Gauchat, "Politicization of Science"; Gauchat, "The Cultural Authority of Science"; Nadelson et al., "I Just Don't Trust Them"; Gauchat, "A Test of Three Theories."
16 Besley and Shanahan, "Media Attention and Exposure."
17 Sarathchandra and Haltinner, "Trust/Distrust Judgments."
18 Sarathchandra and Haltinner, "Trust/Distrust Judgments."
19 McCright et al., "The Influence of Political Ideology on Trust in Science."
20 Hamilton, Hartter, and Saito, "Trust in Scientists on Climate Change."
21 Malka, Krosnick, and Langer, "The Association of Knowledge."
22 Sarathchandra and Haltinner, "A Survey Instrument," 18.
23 Sarathchandra, Haltinner, and Grindal, "Climate Skeptics' Identity Construction"; McCright and Dunlap, "The Politicization of Climate Change."
24 McCright and Dunlap, "The Politicization of Climate Change"; Lahsen, "Anatomy of Dissent"; Elsasser and Dunlap, "Leading Voices in the Denier Choir."
25 Drummond and Fischhoff, "Individuals with Greater Science Literacy."
26 Cooper, "Commercialization of the University"; Stuart and Ding, "When Do Scientists Become Entrepreneurs?"; Azoulay, Ding, and Stuart, "The Impact of Academic Patenting."
27 For example, in 2019 a scandal broke at the prestigious Memorial Sloan Kettering Cancer Center in New York, where a review found top officials at the center had repeatedly violated financial conflicts of interests, which led the center to completely overhaul its policies governing employee relationships with outside companies. This incident received widespread media coverage. See Ornstein and Thomas, "Memorial Sloan Kettering Leaders."
28 Sarathchandra and McCright, "The Effects of Media Coverage."
29 See Watts, "Climatologist Michael E. Mann."
30 Olson et al., "Publication Bias in Editorial Decision Making."
31 Harlos, Edgell, and Hollander, "No Evidence of Publication Bias."
32 Oppenheimer et al., *Discerning Experts.*
33 NOAA, "How Do We Study Past Climates?"
34 For a general and quick overview of climate modeling, see NOAA's webpage on climate models: https://www.climate.gov/maps-data/primer/climate-models.
35 Drummond and Fischhoff, "Individuals with Greater Science Literacy."
36 Kahan et al., "Culture and Identity-Protective Cognition."
37 Fiebrich, "History of Surface Weather Observations."
38 NOAA, "Climate Models."
39 For an overview of the general peer review process and its limitations, see Mayden, "Peer Review," 117.
40 McCright and Dunlap, "The Politicization of Climate Change"; Haltinner and Sarathchandra, "Adding the Church of Jesus Christ of Latter-day Saints."

THREE *Religious Ideation and Conspiracy Adherence*

1. Uscinski, Douglas, and Lewandowsky, "Climate Change Conspiracy Theories"; Lewandowsky, Oberauer, and Gignac, "NASA Faked the Moon Landing"; Lewandowsky, Gignac, and Oberauer, "The Robust Relationship."
2. Althusser, *Lenin and Philosophy and Other Essays.*
3. Žižek, *The Sublime Object of Ideology.*
4. Arendt, "Ideology and Terror."
5. Wang and Kim, "Analysis of the Impact of Values."
6. Foucault, *Discipline and Punish.*
7. Pearce and Thornton, "Religious Identity and Family Ideologies."
8. McCright and Dunlap, "Cool Dudes"; Wang and Kim, "Analysis of the Impact of Values," 99.
9. Haltinner and Sarathchandra, "Beyond Religiosity."
10. Bourdieu, "Rethinking the State."
11. Haltinner and Sarathchandra, "Pro-Environmental Views of Climate Skeptics."
12. Hayhoe, "Kathrine Hayhoe: Climate Scientist."
13. Kearns, "Noah's Ark Goes to Washington."
14. Mahaffy, "Cognitive Dissonance and Its Resolution."
15. White, "The Historical Roots of Our Ecologic Crisis"; Eckberg and Blocker, "Christianity, Environmentalism."
16. Truelove and Joireman, "Understanding the Relationship."
17. Eckberg and Blocker, "Christianity, Environmentalism"; Sherkat and Ellison, "Structuring the Religion-Environment Connection."
18. Haluza-DeLay, "Religion and Climate Change."
19. Haltinner and Sarathchandra, "Beyond Religiosity."
20. Sarathchandra and Haltinner, "How Believing Climate Change Is a Hoax."
21. Lewandowsky and Cook, *The Conspiracy Theory Handbook.*
22. Haltinner, Sarathchandra, and Ptak, "How Believing."
23. Swami et al., "Conspiracist Ideation; Abalakina-Paap et al., "Beliefs in Conspiracies."
24. Swami and Furnham, "Political Paranoia and Conspiracy Theories."
25. Sarathchandra and Haltinner, "How Believing Climate Change Is a Hoax."
26. Sarathchandra and Haltinner, "How Believing Climate Change Is a Hoax."
27. Haltinner, Sarathchandra, and Ptak, "How Believing."
28. Newheiser, Farias, and Tausch, "The Functional Nature of Conspiracy Beliefs"; Galliford and Furnham, "Individual Difference Factors."
29. Sarathchandra and Haltinner, "How Believing Climate Change Is a Hoax."
30. Haltinner, Sarathchandra, and Ptak, "How Believing."
31. Foucault, *Society Must Be Defended.*

32 Northey et al., "LGBTQ Imagery in Advertising."
33 Haltinner and Saratchandra, "Beyond Religiosity."
34 Sarathchandra and Haltinner, "How Believing Climate Change Is a Hoax," 9.
35 Sarathchandra and Haltinner, 9.
36 Haltinner and Sarathchandra, "Beyond Religiosity."
37 Sarathchandra and Haltinner, "How Believing Climate Change Is a Hoax," 9.
38 Haltinner, Ladino, and Sarathchandra, "Feeling Skeptical."
39 Sarathchandra and Haltinner, "How Believing Climate Change Is a Hoax."
40 Goldberg et al, "A Social Identity Approach."
41 Nisbet, "Communicating Climate Change."

FOUR *Pro-Environmentalism among Skeptics*

1 Schulman, "Every Insane Thing"; Popovich, Albeck-Ripka, and Pierre-Louis, "The Trump Administration Is Reversing."
2 Bump, "This Is Perhaps Trump's Most Cynical Comment."
3 Ferretti and Mauger, "Harris-Pence Debate."
4 Kessler, Rizzo, and Kelly, "President Trump Made 16,241 False or Misleading Claims."
5 Saad, "Half in U.S. Are Now Concerned."
6 Haltinner and Sarathchandra, "Predictors of Pro-Environmental Beliefs."
7 EPA, "Our Nation's Air."
8 Mackenzie, "Air Pollution."
9 Mackenzie, "Air Pollution."
10 Haltinner and Sarathchandra, "Predictors of Pro-Environmental Beliefs," 9–10.
11 Taylor, *Nor Any Drop to Drink.*
12 CDC, "Health Effects of Lead Exposure."
13 Denchak, "Flint Water Crisis."
14 Langin, "Millions of Americans Drink."
15 Idaho Department of Agriculture, "Always Growing."
16 NRCS Idaho, "Soil Health."
17 Kinney et al., "Presence and Distribution of Wastewater-Derived Pharmaceuticals."
18 Idaho Department of Environmental Quality, "Safe Pharmaceuticals Disposal."
19 The Ocean Cleanup, "The Great Pacific Garbage Patch."
20 NASA, "Global Climate Change."
21 Sneed, "Ask the Experts."
22 Deng, "In China, the Water You Drink."
23 Lowe, *Immigrant Acts*; Lee, *The Making of Asian America.*
24 De Leon, "The Long History of Racism"; Lee, *The Making of Asian America.*

25 L. Zhou, "Trump's Racist References"; Liu, "The Coronavirus and the Long History."
26 Haltinner and Sarathchandra, "Predictors of Pro-Environmental Beliefs," 9–10.
27 Ceballos et al., "Accelerated Modern Human–Induced Species Losses."
28 Tarlach, "The Five Mass Extinctions."
29 Gerretsen, "One Million Species Threatened."
30 Briggs, "Emergence of a Sixth Mass Extinction?"
31 Carrington, "Earth's Sixth Mass Extinction."
32 Zielinski, "Climate Change Will Accelerate."
33 National Parks Conservation Association, "New Poll of Likely Voters."
34 Haltinner and Sarathchandra, "Predictors of Pro-Environmental Beliefs," 9–10.
35 Union of Concerned Scientists, "Coal Power Impacts."
36 Union of Concerned Scientists, "Coal Power Impacts."
37 Union of Concerned Scientists, "Coal Power Impacts."
38 Haltinner and Sarathchandra, "Predictors of Pro-Environmental Beliefs," 9–10.
39 Haltinner and Sarathchandra, 9–10.
40 Solar Reviews, "How Much Do Solar Panels Cost"; Coren, "The Median Electric Car."
41 The White House, "Fact Sheet."
42 The White House, "The Build Back Better Framework."
43 Ellsmoor, "United States Spend Ten Times More."
44 Ellsmoor, "United States Spend."
45 Hudson, "Energy Independence Is a Farce."
46 Keller, "America's Energy Independence."
47 US Energy Information Administration, "Fossil Fuels Continue to Account."
48 Haltinner and Sarathchandra, "Predictors of Pro-Environmental Beliefs," 13.
49 Haltinner and Sarathchandra, 13.
50 Eagles and Damare, "Factors Influencing Children's Environmental Attitudes"; Hoffmann et al., "Climate Change Experiences."
51 Arp and Kenny, "Black Environmentalism"; Gifford and Nilsson, "Personal and Social Factors."
52 Gifford and Nilsson, "Personal and Social Factors."
53 Prati and Zani, "The Effect of the Fukushima."
54 Wachinger et al., "The Risk Perception Paradox."
55 Rees, Klug, and Bamberg, "Guilty Conscience"; Li, "Nostalgia Promoting Pro-Social Behavior."
56 Schwartz and Loewenstein, "The Chill of the Moment."
57 Haltinner, Ladino, and Sarathchandra, "Feeling Skeptical."
58 Haltinner, Ladino, and Sarathchandra, "Feeling Skeptical."
59 Gallese, "The Manifold Nature."

60 Wachinger et al., "The Risk Perception Paradox," 1060.
61 Shan, "Can Virtual Reality Drive Sustainable Behavior?"
62 Wang et al., "Emotions Predict Policy Support."
63 Pellow and Vazin, "The Intersection of Race."
64 BBC, "Gerard Butler, Miley Cyrus."
65 Leiserowitz, "American Risk Perceptions."
66 UN Environment Program, "Air Pollution and Climate Change."
67 CDC, "Climate Change Decreases the Quality of the Air We Breathe."
68 Dunne, "Deforestation Has Driven Up"; Union of Concerned Scientists, "Benefits of Renewable Energy Use."
69 NREL, "Renewable Electricity Futures Study," 210.
70 NREL, "Renewable Electricity Futures Study."

FIVE *Engagement with Media and Information*

1 Alamassi, "Climate Change, Epistemic Trust."
2 Brulle, Carmichael, and Jenkins, "Shifting Public Opinion."
3 Brulle, Carmichael, and Jenkins, "Shifting Public Opinion."
4 Dunlap and McCright, "Climate Change Denial."
5 Boykoff, "Public Enemy No. 1?"
6 Boykoff, "Public Enemy No. 1?"
7 Boykoff and Boykoff, "Balance as Bias."
8 Schmid-Petri et al., "A Changing Climate of Skepticism"; Painter and Ashe, "Cross-National Comparison."
9 Harvey et al., "Internet Blogs, Polar Bears."
10 Nisbet, "Communicating Climate Change."
11 Sarathchandra and Ten Eyck, "To Tell the Truth."
12 Nisbet, "Communicating Climate Change."
13 Dixon, Hmielowski, and Ma, "Improving Climate Change Acceptance."
14 Hart and Nisbet, "Boomerang Effects."
15 Zhou, "Boomerangs versus Javelins"; Carmichael, Brulle, and Huxster, "The Great Divide."
16 Dunlap, "Climate Change Skepticism and Denial."
17 Dunlap and Brulle, "Sources and Amplifiers of Climate Change Denial."
18 Brulle, "Networks of Opposition."
19 Powell, *The Inquisition of Climate Science*; Oreskes and Conway, *Merchants of Doubt.*
20 Dunlap and McCright, "Climate Change Denial."
21 Hmielowski et al., "An Attack on Science?"
22 Dunlap, "Climate Change Skepticism and Denial."

23 Farrell, "Corporate Funding and Ideological Polarization."
24 Farrell, "The Growth of Climate Change Misinformation."
25 Farrell, "The Growth of Climate Change Misinformation."
26 This potentially reflects the low popularity of these media outlets around the time our survey was conducted in late 2019 and early 2020. For example, One America News (OAN) increased its online engagement significantly leading up to and right after the November 2020 US presidential election, which occurred after our data collection. It is also plausible that skeptics who engaged with fringe media outlets at the time were seeking information on other, more politically charged topics that from their standpoint were seen as more urgent and timely.
27 Eichler, "Glenn Beck, Climate Change Believer."
28 Bruns, "Filter Bubble."
29 Coaston, "YouTube, Facebook, and Apple's Ban."
30 Del Vicario et al., "Modeling Confirmation Bias and Polarization."
31 See links to Michael Savage's books and articles from the aggregate site SourceWatch here: https://www.sourcewatch.org/index.php/Michael_Savage.
32 Elssaser and Dunlap, "Leading Voices in the Denier Choir."
33 Jasny and Fisher, "Echo Chambers in Climate Science."
34 Zhou, "Boomerangs versus Javelins."
35 Sarathchandra and Haltinner, "How Climate Skeptics' Information Sources."
36 Dunlap and McCright, "Challenging Climate Change."
37 Mullainathan and Shleifer, "Media Bias."
38 Pereira and Van Bavel, "Identity Concerns Drive Belief."
39 McCright and Dunlap, "The Politicization of Climate Change."
40 Boykoff and Boykoff, "Balance as Bias."
41 Boykoff and Boykoff, "Balance as Bias."
42 Quandt et al., "Fake News."
43 Zhou, "Boomerangs versus Javelins."
44 Boykoff and Boykoff, "Balance as Bias."
45 Cook, "Deconstructing Climate Science Denial."
46 Lewandowsky, "Climate Change Disinformation."
47 See Cook, "Deconstructing Climate Science Denial," for a full list of climate denial techniques.
48 Taber and Lodge, "Motivated Skepticism."
49 Flaxman, Goel, and Rao, "Filter Bubbles, Echo Chambers."
50 Pereira and Van Bavel, "Identity Concerns Drive Belief."
51 Nisbet, "Communicating Climate Change."
52 Smith, "Republicans Who Believe in Climate Change."
53 Brady, "'Light Years Ahead."

SIX *The Emotional Lives of Skeptics*

1 Cli-fi (climate fiction) is a growing subgroup of science fiction literature that deals with the impacts of climate change on environment and society.

2 Yale Climate Opinion Maps, 2021.

3 Sobel, "Climate Change Meets Ecophobia"; Albrecht, *Earth Emotions*; Haltinner, Ladino, and Sarathchandra, "Feeling Skeptical."

4 Chapman, Lickel, and Markowitz, "Reassessing Emotion."

5 Slovic et al., "The Affect Heuristic."

6 Epstein, "Integration of the Cognitive."

7 Ring, "Inspire Hope, Not Fear."

8 Norgaard, *Living in Denial.*

9 Randall, "Loss and Climate Change."

10 Roberts, "Did that New York Magazine."

11 Haltinner and Sarathchandra, "Climate Change Skepticism."

12 Chapman, Lickel, and Markowitz, "Reassessing Emotion."

13 Haltinner and Sarathchandra, "Climate Change Skepticism

14 Haltinner, Ladino, and Sarathchandra, "Feeling Skeptical."

15 Ray, *A Field Guide to Climate Anxiety.*

16 Haltinner, Ladino, and Sarathchandra, "Feeling Skeptical," 3.

17 Haltinner, Ladino, and Sarathchandra, "Feeling Skeptical," 3.

18 McCright and Dunlap, "Bringing Ideology In.".

19 Smith and Leiserowitz, "The Role of Emotion."

20 Ring, "Inspire Hope, Not Fear."

21 Stanley et al., "From Anger to Action."

22 Stanley et al., "From Anger to Action."

23 Myers et al., "A Public Health Frame."

24 Yeuh, "Classification of Conversations."

25 McCall and Simmons, *Identities and Interactions.*

26 Stryker, "The Interplay of Affect and Identity."

27 Stets and Tsushima, "Negative Emotion and Coping Responses."

28 Ellestad and Stets, "Jealousy and Parenting."

29 Messner, *Politics of Masculinities.*

30 Kimmel, *Angry White Men.*

31 See chapter 3 for more details on skeptics' ideological beliefs, including conspiracy ideation.

32 Chapter 5 further explores skeptics' distrust of media, including mainstream media.

33 Feldman and Hart, "Is There Any Hope?"

34 Du Bray et al., "Anger and Sadness."

35 Haltinner, Ladino, and Sarathchandra, "Feeling Skeptical."
36 Haltinner, Ladino, and Sarathchandra, "Feeling Skeptical."
37 Haltinner, Ladino, and Sarathchandra, "Feeling Skeptical."
38 Haltinner, Ladino, and Sarathchandra, "Feeling Skeptical."
39 Kimmel, "Angry White Men."
40 Haltinner, Ladino, and Sarathchandra, "Feeling Skeptical."
41 Haltinner and Sarathchandra, "Climate Change Skepticism."
42 Krause and Hayward, "Emotional Expressiveness."
43 Krause and Hayward, "Emotional Expressiveness."
44 Giffords and Nilsson, "Personal and Social Factors."
45 For the composite measures used to quantify environmental concern and policy support, see Haltinner and Sarathchandra, "Predictors of Pro-Environmental Beliefs."
46 Wang et al., "Emotions Predict Policy Support."
47 Haltinner, Ladino, and Sarathchandra, "Feeling Skeptical."
48 Lohan, "Can California's Iconic Redwoods Survive?"
49 Ring, "Inspire Hope, Not Fear."

SEVEN *Toward a Continuum of Skepticism*

1 Poortinga et al., "Uncertain Climate"; Capstick and Pidgeon, "What Is Climate Change Scepticism?"
2 Poortinga et al., "Uncertain Climate"; McCright and Dunlap, "Anti-Reflexivity"; Rahmstorf, "The Climate Sceptics."
3 Dunlap, "Climate Change Skepticism and Denial."
4 Haltinner and Sarathchandra, "Considering Attitudinal Uncertainty," 102243.
5 Haltinner and Sarathchandra, "Considering Attitudinal Uncertainty."
6 Dunlap, "Climate Change Skepticism and Denial."
7 Yale Climate Opinion Maps, 2021.
8 Borick et al., "As Americans Experienced."
9 Poortinga et al., "Uncertain Climate."
10 Poortinga et al., "Uncertain Climate."
11 Poortinga et al., "Uncertain Climate." For more information about the denial machine, see Dunlap and McCright, "Challenging Climate Change."
12 In creating this name we borrow from Capstick and Pidgeon's reclassification of skeptics in "What Is Climate Change Scepticism?," which was written in response to Rahmstorf's "The Climate Sceptics." Capstick and Pidgeon find significant overlap between trend skeptics and those who also show signs of attribution, impact, and consensus skepticism. In line with our argument that skepticism should be thought of as an umbrella term, we apply the terminology of denier here.

13 Here we borrow and modify Rahmstorf's term. See, Rahmstorf, "The Climate Sceptics."

14 See chapter 3 for more discussion on conspiracy theories and climate change. Also see Sarathchandra and Haltinner, "How Believing Climate Change Is a 'Hoax.'"

15 Rahmstorf, "The Climate Sceptics."

16 McCright and Dunlap, "Cool Dudes."

17 See appendix, table 3, for a comparison of these climate change beliefs across the skepticism continuum.

18 Appendix, table 3.

19 Appendix, table 3.

20 Appendix, table 3.

21 Appendix, table 3.

22 Appendix, table 3.

23 Haltinner and Sarathchandra, "Considering Attitudinal Uncertainty."

24 Appendix, table 3.

25 Haltinner, Ladino, and Sarathchandra, "Feeling Skeptical."

26 Leiserowitz et al., "Climate Change in the American Mind."

27 See appendix, table 2.

28 Appendix, table 2.

29 Haltinner and Sarathchandra, "Considering Attitudinal Uncertainty," 5; Appendix, table 3.

30 See appendix, table 3.

31 Appendix, table 3.

32 Appendix, table 2.

33 Appendix, table 3.

34 Appendix, table 3.

35 Appendix, table 3.

36 Haltinner and Sarathchandra, "Considering Attitudinal Uncertainty," 5.

37 Appendix, table 3.

38 Haltinner and Sarathchandra, "Climate Change Skepticism."

39 Appendix, table 2.

40 Appendix, table 2.

41 Appendix, table 3.

42 Appendix, table 3.

43 Appendix, table 3.

44 Appendix, table 3.

45 Appendix, table 3.

46 Mean difference=1.20; $p<.001$ (trend); mean difference=0.78; $p<.001$ (attribution); mean difference=1.27; $p<.001$ (impact); mean difference=1.08; $p<.001$ (consensus).

47 Mean difference=1.08; $p<.000$.
48 Chi square=296.34; $p<.000$.
49 Shipman and Kay, *Womenomics.*
50 Stets and Burke, "A Sociological Approach."
51 Dunlap, "Climate Change Skepticism and Denial."

EIGHT *Changing Perceptions of Climate Change*

1 Yale Climate Opinion Maps, 2021.
2 Ballew et al., "Climate Change in the American Mind."
3 Motta, "Changing Minds or Changing Samples?"
4 Motta, "Changing Minds or Changing Samples?"
5 Harvey, "World Is Running Out of Time."
6 Nisbet, "Communicating Climate Change"; Suldovsky, "The Information Deficit Model."
7 The data presented here comes from the second round of interviews, conducted in 2018–2020. Haltinner and Sarathchandra, "It Wasn't Like a Big Lightbulb Moment."
8 Mousa, "Building Social Cohesion."
9 Suldovsky, "The Information Deficit Model."
10 Burke, "Identity Change."
11 Burke, "Identity Change."
12 Carmichael, Brulle, and Huxster, "The Great Divide."
13 Van Bavel and Packer, *The Power of Us.*
14 Anderson, Pender, and Asner-Self, "A Review of the Religious Identity/Sexual Orientation."
15 McKimmie et al., "I'm a Hypocrite."
16 Turner et al., *Rediscovering the Social Group.*
17 McKimmie et al., "I'm a Hypocrite," 214.
18 Mahaffy, "Cognitive Dissonance."
19 Mahaffy, "Cognitive Dissonance."
20 Bavel and Packer, *The Power of Us.*
21 Arp and Kenny, "Black Environmentalism"; Gifford and Nilsson, "Personal and Social Factors."
22 Prati and Zani, "The Effect of the Fukushima."
23 Eagles and Damare, "Factors Influencing Children's Environmental Attitudes"; Rosa, Profice, and Collado, "Nature Experiences"; Hoffmann et al., "Climate Change Experiences."
24 Haltinner and Sarathchandra, "Predictors of Pro-Environmental Beliefs."
25 Arceneaux and Johnson, *Changing Minds or Changing Channels?*

26 Manjoo, *True Enough.*
27 Markowitz and Shariff, "Climate Change and Moral Judgement."
28 Carmichael, Brulle, and Huxster, "The Great Divide."
29 Suldovsky, "The Information Deficit Model."
30 Boykoff, *Who Speaks for the Climate?*
31 McDonald, Chai, and Newell, "Personal Experience."
32 Joiremann, Truelove, and Duell, "Effect of Outdoor Temperature"; Egan and Mullin, "Turning Personal Experience."

CONCLUSION

1 Melley, "Explosive California Wildfires"; Flood List, "Philippines"; "North China Floods"; France24 Newswire, "Floods and Landslides."
2 McGrath, "Climate Change."
3 Price, "IPCC Report."
4 Yale Climate Opinion Maps, 2021.
5 Kahan et al., "Culture and Identity-Protective Cognition."
6 Ring, "Inspire Hope, Not Fear."
7 Lewandowsky and Cook, *The Conspiracy Theory Handbook.*
8 RepublicEn, "Are You One of Us?"
9 Hestres, "Fighting Climate Change Denial."
10 Haltinner and Sarathchandra, "Predictors of Pro-Environmental Beliefs."
11 American Conservation Coalition (ACC) is a 501(c)(4) nonprofit organization mobilizing young conservatives toward climate and environmental action. Learn more at https://www.acc.eco/about-acc. The Green New Deal is a congressional resolution stemming from the US political left, which targets mobilization of the American society toward renewable energy as well as social and economic justice. Learn more at https://www.congress.gov/bill/116th-congress/house-resolution/109/text. The American Climate Contract is presented by young conservative climate activists as a response to the Green New Deal. Learn more at https://www.climatesolution.eco.
12 Matthews, "ACC Campus President."
13 Young Conservatives for Carbon Dividends categorizes itself as a free-market climate advocacy campaign. Learn more at https://www.yccdaction.org.
14 Haltinner and Sarathchandra, "Adding the Church of Jesus Christ of Latter-day Saints."
15 Roeser, "Risk Communication, Public Engagement."
16 Haltinner and Sarathchandra, "Adding the Church of Jesus Christ of Latter-day Saints."
17 Haltinner and Sarathchandra, "Climate Change Skepticism."

18 Chapman, Lickel, and Markowitz, "Reassessing Emotion."

19 Haltinner and Sarathchandra, "Predictors of Pro-Environmental Beliefs."

20 Haltinner and Sarathchandra, "Pro-Environmental Views."

21 Ring, "Inspire Hope, Not Fear."

22 See the National Academy of Sciences Communication Collection at https://www.nap.edu/collection/68/science-communication.

23 Hornsey and Fielding, "Attitude Roots and Jiu Jitsu Persuasion," 459.

24 Haltinner and Sarathchandra, "Pro-Environmental Views."

25 Alrababa'h et al., "Can Exposure to Celebrities Reduce Prejudice?"

26 Schutte and Stilinović, "Facilitating Empathy," 708–712; Johnson, "Transportation into a Story"; Hasler et al., "Virtual Peacemakers"; Johnson, "Transportation into Literary Fiction"; Taylor and Glen, "From Empathy to Action."

27 Lewandowsky and Cook, *The Conspiracy Theory Handbook.*

28 Van Bavel, "Group Think."

APPENDIX

1 Pilgeram, *Pushed Out.*

2 Glaser, *The Grounded Theory Perspective*; Weed, "Capturing the Essence of Grounded Theory."

bibliography

Abalakina-Paap, Marina, Walter G. Stephan, Traci Craig, and W. Larry Gregory. "Beliefs in Conspiracies." *Political Psychology* 20, no. 3 (1999): 637–47.

Albrecht, Glenn. *Earth Emotions: New Words for a New World*. Ithaca, NY: Cornell University Press, 2019.

Allport, Gordon W. "The Historical Background of Social Psychology." In *The Handbook of Social Psychology, 3rd ed.*, edited by Gardner Lindzey and Elliot Aronson, 1–46. New York: Random House, 1985.

Almassi, Ben. "Climate Change, Epistemic Trust, and Expert Trustworthiness." *Ethics & the Environment* 17, no. 2 (2012): 29–49.

Alrababa'h, Ala', William Marble, Salma Mousa, and Alexandra Siegel. "Can Exposure to Celebrities Reduce Prejudice? The Effect of Mohamed Salah on Islamophobic Behaviors and Attitudes." *American Political Science Review* (2021): 1–18.

Althusser, Louis. *Lenin and Philosophy and Other Essays*. New York: Aakar, 2006.

American Academy of Arts and Sciences. "Perceptions of Science in America." 2018. https://www.amacad.org/publication/perceptions-science-america.

Anderton, Cindy L., Debra A. Pender, and Kimberly K. Asner-Self. "A Review of the Religious Identity/Sexual Orientation Identity Conflict Literature: Revisiting Festinger's Cognitive Dissonance Theory." *Journal of LGBT Issues in Counseling* 5, no. 3–4 (2011): 259–81.

Arceneaux, Kevin, and Martin Johnson. *Changing Minds or Changing Channels? Partisan News in an Age of Choice*. Chicago: University of Chicago Press, 2013.

Arendt, Hannah. "Ideology and Terror: A Novel Form of Government." *Review of Politics* 15, no. 3 (1953): 303–27.

Arp, Wiwam, and Christopher Kenny. "Black Environmentalism in the Local Community Context." *Environment and Behavior* 28 (1996): 267–82.

Azoulay, Pierre, Waverly Ding, and Toby Stuart. "The Impact of Academic Patenting on the Rate, Quality and Direction of (Public) Research Output." *Journal of Industrial Economics* 57, no. 4 (2009): 637–76.

Ballew, Matthew T., Anthony Leiserowitz, Connie Roser-Renouf, Seth A. Rosenthal, John E. Kotcher, Jennifer R. Marlon, Erik Lyon, Matthew H. Goldberg, and

Edward W. Maibach. "Climate Change in the American Mind: Data, Tools, and Trends." *Environment: Science and Policy for Sustainable Development* 61, no. 3 (2019): 4–18.

Besley, John C., and James Shanahan. "Media Attention and Exposure in Relation to Support for Agricultural Biotechnology." *Science Communication* 26, no. 4 (2005): 347–67.

Bliuc, Ana-Maria, Craig McGarty, Emma F. Thomas, Girish Lala, Mariette Berndsen, and RoseAnne Misajon. "Public Division about Climate Change Rooted in Conflicting Socio-Political Identities." *Nature Climate Change* 5, no. 3 (2015): 226–29.

Boaz, David. *Libertarianism*. New York: Free Press, 1997.

Borick, Christopher, Barry G. Rabe, Natalie B. Fitzpatrick, and S. B. Mills. "As Americans Experienced the Warmest May on Record Their Acceptance of Global Warming Reaches a New High." *Issues in Energy and Environmental Policy (IEEP)* 37 (2018).

Bourdieu, Pierre. "Rethinking the State: Genesis and Structure of the Bureaucratic Field." In *State/Culture*, edited by George Steinmetz, 53–75. Ithaca, NY: Cornell University Press, 2018.

Brady, Jeff. "Light Years Ahead of Their Elders, Young Republicans Push GOP on Climate Change." National Public Radio, September 25, 2020. https://www.npr.org/2020/09/25/916238283/light-years-ahead-of-their-elders-young-republicans-push-gop-on-climate-change.

Branscombe, Nyla R., Naomi Ellemers, Russell Spears, and Bertjan Doosje. "The Context and Content of Social Identity Threat." *Social Identity: Context, Commitment, Content* 1 (1999): 35–58.

Branscombe, Nyla R., Michael T. Schmitt, and Richard D. Harvey. "Perceiving Pervasive Discrimination among African Americans: Implications for Group Identification and Well-Being." *Journal of Personality and Social Psychology* 77, no. 1 (1999): 135.

Boykoff, Maxwell T. "Public Enemy No. 1? Understanding Media Representations of Outlier Views on Climate Change." *American Behavioral Scientist* 57, no. 6 (2013): 796–817.

Boykoff, Maxwell T. *Who Speaks for the Climate? Making Sense of Media Reporting on Climate Change*. Cambridge, MA: Cambridge University Press, 2011.

Boykoff, Maxwell T., and Jules M. Boykoff. "Balance as Bias: Global Warming and the US Prestige Press." *Global Environmental Change* 14, no. 2 (2004): 125–36.

Brewer, Marilynn B. "The Psychology of Prejudice: Ingroup Love or Outgroup Hate?" *Journal of Social Issues* 55 (1999): 429–44.

Brewer, Paul R., and Barbara L. Ley. "Whose Science Do You Believe? Explaining Trust in Sources of Scientific Information about the Environment." *Science Communication* 35, no. 1 (2013): 115–37.

Briggs, John C. "Emergence of a Sixth Mass Extinction?" *Biological Journal of the Linnean Society* 122, no. 2 (2017): 243–48.

British Broadcasting Corporation. "Gerard Butler, Miley Cyrus: Stars' Homes Destroyed by California Wildfires." *BBC News*, October 12, 2020. https://www.bbc.com/news/entertainment-arts-46178520.

Brodsky, Angela N. "Constructing Deniers: Identity, Discourse, and Stigma Negotiation in the Climate Change Debate on Yahoo! Comment Forums." PhD diss., University of West Georgia, 2015.

Brulle, Robert J. "Networks of Opposition: A Structural Analysis of US Climate Change Countermovement Coalitions 1989–2015." *Sociological Inquiry* 91, no. 3 (2021): 603–24.

Brulle, Robert J., Jason Carmichael, and J. Craig Jenkins. "Shifting Public Opinion on Climate Change: An Empirical Assessment of Factors Influencing Concern over Climate Change in the US, 2002–2010." *Climatic Change* 114, no. 2 (2012): 169–88.

Bruns, Axel. "Filter Bubble." *Internet Policy Review* 8, no. 4 (2019). https://doi.org/10.14763/2019.4.1426.

Bugden, Dylan. "Denial and Distrust: Explaining the Partisan Climate Gap." *Climatic Change* 170, no. 3 (2022): 1–23.

Bump, Philip. "This Is Perhaps Trump's Most Cynical Comment about the Environment Yet." *Washington Post,* January 9, 2020. https://www.washingtonpost.com/politics/2020/01/09/this-is-perhaps-trumps-most-cynical-comment-about-environment-yet/.

Burke, Peter J. "Identity Change." *Social Psychology Quarterly* 69, no. 1 (2006): 81–96.

Burke, Peter J., and Jan E. Stets. *Identity Theory*. Oxford, UK: Oxford University Press, 2009.

Burke, Peter J., and Judy Tully. "The Measurement of Role/Identity." *Social Forces* 55 (1977): 347–66.

Carmichael, Jason T., Robert J. Brulle, and Joanna K. Huxster. "The Great Divide: Understanding the Role of Media and Other Drivers of the Partisan Divide in Public Concern Over Climate Change in the USA, 2001–2014." *Climatic Change* 141, no. 4 (2017): 599–612.

Carrington, Damian. "Earth's Sixth Mass Extinction Event Under Way, Scientists Warn." *Guardian,* July 10, 2017. https://www.theguardian.com/environment/2017/jul/10/earths-sixth-mass-extinction-event-already-underway-scientists-warn.

CDC (Centers for Disease Control and Prevention). "Climate Change Decreases the Quality of the Air We Breathe." Accessed October 10, 2020. https://www.cdc.gov/climateandhealth/pubs/air-quality-final_508.pdf.

Ceballos, Gerardo, Paul R. Ehrlich, Anthony D. Barnosky, Andrés García, Robert M. Pringle, and Todd M. Palmer. "Accelerated Modern Human-Induced Species Losses: Entering the Sixth Mass Extinction." *Science Advances* 1, no. 5 (2015). https://doi.org/10.1126/sciadv.1400253.

Center for the Study of the Pacific Northwest. "Lesson One: Who Belongs in the Pacific Northwest." Accessed March 21, 2020. https://www.washington.edu/uwired/outreach/cspn/Website /Classroom%20Materials/Pacific%20Northwest%20History/Lessons/Lesson%201/1.html.

Center for the Study of the Pacific Northwest. "Lesson Fourteen: Industrialization, Technology, and Environment in Washington." Accessed November 14, 2021. https://www.washington.edu/uwired/outreach/cspn/Website/Classroom%20Materials/Pacific%20Northwest%20History/Lessons/Lesson%2014/14.html. Website discontinued.

CDC (Centers for Disease Control and Prevention). "Health Effects of Lead Exposure." Accessed January 7, 2020. https://www.cdc.gov/nceh/lead/prevention/health-effects.htm.

Chapman, Daniel A., Brian Lickel, and Ezra M. Markowitz. "Reassessing Emotion in Climate Change Communication." *Nature Climate Change* 7, no. 12 (2017): 850–52.

Coaston, Jane. "YouTube, Facebook, and Apple's Ban on Alex Jones, Explained." Vox, August 6, 2018. https://www.vox.com/2018/8/6/17655658/alex-jones-facebook-youtube-conspiracy-theories.

Cook, John. "Deconstructing Climate Science Denial." In *Research Handbook on Communicating Climate Change*, edited by David C. Holmes and Lucy M. Richardson, 62–78. Cheltenham, UK: Edward Elgar, 2020.

Cooper, Mark H. "Commercialization of the University and Problem Choice by Academic Biological Scientists." *Science, Technology, & Human Values* 34, no. 5 (2009): 629–53.

Capstick, Stuart Bryce, and Nicholas Frank Pidgeon. "What Is Climate Change Scepticism? Examination of the Concept Using a Mixed Methods Study of the UK Public." *Global Environmental Change* 24 (2014): 389–401.

Coren, Michael. "The Median Electric Car in the US Is Getting Cheaper." *Quartz*, August 26, 2019. https://qz.com/1695602/the-average-electric-vehicle-is-getting-cheaper-in-the-us/.

De Leon, Adrian. "The Long History of Racism against Asian Americans in the U.S." Public Broadcasting System. Accessed April 9, 2020. https://www.pbs.org/newshour/nation/the-long-history-of-racism-against-asian-americans-in-the-u-s.

Dietrich, William. *The Final Forest: Big Trees, Forks, and the Pacific Northwest*. Seattle: University of Washington Press, 2011.

Del Vicario, Michela, Antonio Scala, Guido Caldarelli, H. Eugene Stanley, and Walter Quattrociocchi. "Modeling Confirmation Bias and Polarization." *Scientific Reports* 7, no. 1 (2017): 1–9.

Denchak, Melissa. "Flint Water Crisis: Everything You Need to Know." National Resources Defense Council. Accessed November 8, 2018. https://www.nrdc.org/stories/flint-water-crisis-everything-you-need-know.

Dixon, Graham, Jay Hmielowski, and Yanni Ma. "Improving Climate Change Acceptance among US Conservatives through Value-Based Message Targeting." *Science Communication* 39, no. 4 (2017): 520–34.

Du Bray, Margaret, Amber Wutich, Kelli L. Larson, Dave D. White, and Alexandra Brewis. "Anger and Sadness: Gendered Emotional Responses to Climate Threats in Four Island Nations." *Cross-Cultural Research* 53, no. 1 (2019): 58–86.

Drummond, Caitlin, and Baruch Fischhoff. "Individuals with Greater Science Literacy and Education Have More Polarized Beliefs on Controversial Science Topics." *Proceedings of the National Academy of Sciences* 114, no. 36 (2017): 9587–92.

Dunlap, Riley E. "Climate Change Skepticism and Denial: An Introduction." *American Behavioral Scientist* 57, no. 6 (2013): 691–98.

Dunlap, Riley E., and Robert Brulle. "Sources and Amplifiers of Climate Change Denial." In *Research Handbook on Communicating Climate Change*, edited by D. C. Holmes and L. M. Richardson, 49–61. Cheltenham: Edward Elgar, 2020.

Dunlap, Riley E., and Peter J. Jacques. "Climate Change Denial Books and Conservative Think Tanks: Exploring the Connection." *American Behavioral Scientist* 57, no. 6 (2013): 699–731.

Dunlap, Riley E., and Aaron M. McCright. "Challenging Climate Change." In *Climate Change and Society: Sociological Perspectives*, edited by Riley E. Dunlap and Aaron M. McCright, 300–332. Oxford, UK: Oxford University Press, 2015.

Dunlap, Riley E., and Aaron M. McCright. "Climate Change Denial: Sources, Actors and Strategies." In *Routledge Handbook of Climate Change and Society*, edited by Constance Lever-Tracy, 270–90. London: Routledge, 2010.

Dunlap, Riley E., and Aaron M. McCright. "Organized Climate Change Denial." In *The Oxford Handbook of Climate Change and Society*, edited by John S. Dryzek, Richard B. Norgaard, and David Schlosberg, 144–60. Oxford, UK: Oxford University Press, 2011.

Dunne, Daisy. "Deforestation Has Driven Up Hottest Day Temperatures, Study Says." Carbon Brief. Accessed October 10, 2020. https://www.carbonbrief.org/deforestation-has-driven-up-hottest-day-temperatures.

Eagles, Paul, and Robert Damare. "Factors Influencing Children's Environmental Attitudes." *Journal of Environmental Education* 30, no. 4 (1999): 33–37.

Eckberg, Douglas Lee, and T. Jean Blocker. "Christianity, Environmentalism, and the Theoretical Problem of Fundamentalism." *Journal for the Scientific Study of Religion* 35, no. 4 (1996): 343–55.

Egan, Patrick J., and Megan Mullin. "Turning Personal Experience into Political Attitudes: The Effect of Local Weather on Americans' Perceptions about Global Warming." *Journal of Politics* 74, no. 3 (2012): 796–809.

Eichler, Alex. "Glenn Beck, Climate Change Believer." *Atlantic*, February 25, 2010. https://www.theatlantic.com/technology/archive/2010/02/glenn-beck-climate-change-believer/341282/.

Ellestad, June, and Jan E. Stets. "Jealousy and Parenting: Predicting Emotions from Identity Theory." *Sociological Perspectives* 41, no. 3 (1998): 639–68.

Ellsmoor, James. "United States Spend Ten Times More on Fossil Fuel Subsidies than Education." *Forbes*, June 15, 2019. https://www.forbes.com/sites/jamesellsmoor/2019/06/15/united-states-spend-ten-times-more-on-fossil-fuel-subsidies-than-education/#280b7bf74473.

Elsasser, Shaun W., and Riley E. Dunlap. "Leading Voices in the Denier Choir: Conservative Columnists' Dismissal of Global Warming and Denigration of Climate Science." *American Behavioral Scientist* 57, no. 6 (2013): 754–76.

Engels, Anita, Otto Hüther, Mike Schäfer, and Hermann Held. "Public Climate-Change Skepticism, Energy Preferences and Political Participation." *Global Environmental Change* 23, no. 5 (2013): 1018–27.

EPA (Environmental Protection Agency). "Our Nation's Air." Accessed March 6, 2020. https://gispub.epa.gov/air/trendsreport/2019/#introduction.

Epstein, Seymour. "Integration of the Cognitive and the Psychodynamic Unconscious." *American Psychologist* 49, no. 8 (1994): 709–24.

Farrell, Justin. "Corporate Funding and Ideological Polarization about Climate Change." *Proceedings of the National Academy of Sciences* 113, no. 1 (2016): 92–97.

Farrell, Justin. "The Growth of Climate Change Misinformation in US Philanthropy: Evidence from Natural Language Processing." *Environmental Research Letters* 14, no. 3 (2019). https://doi.org/10.1088/1748-9326/aaf939.

Feldman, Lauren, and P. Sol Hart. "Is There Any Hope? How Climate Change News Imagery and Text Influence Audience Emotions and Support for Climate Mitigation Policies." *Risk Analysis* 38, no. 3 (2018): 585–602.

Ferretti, Christine, and Craig Mauger. "Harris-Pence Debate: Candidates Spar on COVID-19, Foreign Policy." *Detroit News*, October 7, 2020. https://www.detroitnews.com/story/news/politics/2020/10/07/mike-pence-kamala-harris-vice-presidential-debate-blog/5916571002/.

Fiebrich, Christopher A. "History of Surface Weather Observations in the United States." *Earth-Science Reviews* 93, no. 3–4 (2009): 77–84.

Flaxman, Seth, Sharad Goel, and Justin M. Rao. "Filter Bubbles, Echo Chambers, and Online News Consumption." *Public Opinion Quarterly* 80, no. S1 (2016): 298–320.

Flood List. "Philippines—Tropical Storm Komasu Causes Deadly Flooding." Accessed January 17, 2022. https://floodlist.com/asia/philippines-kompasu-floods-ocotober-2021.

Foucault, Michel. *Discipline and Punish: The Birth of the Prison*. New York: Penguin, 1991.

Foucault, Michel. *Society Must Be Defended: Lectures at the College de France 1975–1976.* New York: Picador, 2003.

France24 Newswire. "Floods and Landslides Leave Dozens Dead in Southwest India." Accessed October 17, 2021. https://www.france24.com/en/asia-pacific/20211017-floods-landslides-leave-dozens-dead-in-southwest-india.

Gallese, V. "The Manifold Nature of Interpersonal Relations: The Quest of a Common Mechanism." *Philosophical Transactions of the Royal Society of London*, series B, no. 358 (2003): 517–28.

Galliford, Natasha, and Adrian Furnham. "Individual Difference Factors and Beliefs in Medical and Political Conspiracy Theories." *Scandinavian Journal of Psychology* 58, no. 5 (2017): 422–28.

Gauchat, Gordon. "The Cultural Authority of Science: Public Trust and Acceptance of Organized Science." *Public Understanding of Science* 20, no. 6 (2011): 751–70.

Gauchat, Gordon. "Politicization of Science in the Public Sphere: A Study of Public Trust in the United States, 1974 to 2010." *American Sociological Review* 77, no. 2 (2012): 167–87.

Gauchat, Gordon. "A Test of Three Theories of Anti-Science Attitudes." *Sociological Focus* 41, no. 4 (2008): 337–57.

Gerretsen, Isabelle. "One Million Species Threatened with Extinction Because of Humans." *CNN,* May 7, 2019. https://www.cnn.com/2019/05/06/world/one-million-species-threatened-extinction-humans-scn-intl/index.html.

Gifford, Robert, and Andreas Nilsson. "Personal and Social Factors that Influence Pro-Environmental Concern and Behavior: A Review." *International Journal of Psychology* 49, no. 3 (2014): 141–57.

Glaser, Barney. *The Grounded Theory Perspective.* Mill Valley, CA: Sociology Press, 2001.

Glaser, Barney, and Anselm Strauss. *The Discovery of Grounded Theory.* New York: Routledge, 2017.

Goffman, Erving. *Stigma: Notes on the Management of Spoiled Identity.* New York: Simon and Schuster, 1963.

Goldberg, Matthew H., Abel Gustafson, Matthew T. Ballew, Seth A. Rosenthal, and Anthony Leiserowitz. "A Social Identity Approach to Engaging Christians in the Issue of Climate Change." *Science Communication* 41 (2019): 442–63.

Hackel, Leor M., Géraldine Coppin, Michael J. A. Wohl, and Jay J. Van Bavel. "From Groups to Grits: Social Identity Shapes Evaluations of Food Pleasantness." *Journal of Experimental Social Psychology* 74 (2018): 270–80.

Haltinner, Kristin. *No Perfect Birth.* London: Rowman and Littlefield, 2021.

Haltinner, Kristin, Jennifer Ladino, and Dilshani Sarathchandra. "Feeling Skeptical: Worry, Dread, and Support for Environmental Policy Among Climate Change Skeptics." *Emotion, Space and Society* 39 (2021). https://doi.org/10.1016/j.emospa.2021.100790.

Haltinner, Kristin, and Dilshani Sarathchandra. "Adding the Church of Jesus Christ of Latter-day Saints to Analyses of Climate Change Skepticism: A Research Note." *Sociological Inquiry* 92, no. 1 (2022): 270–94.

Haltinner, Kristin, and Dilshani Sarathchandra. "Beyond Religiosity: Examining the Relative Effects of Religiosity and Religious Ideation on Climate Skepticism, A Research Note." *Journal of Rural Social Science* 37, no. 3 (2022): 1–29.

Haltinner, Kristin, and Dilshani Sarathchandra. "Climate Change Skepticism as a Psychological Coping Strategy." *Sociology Compass* 12, no. 6 (2018): e12586.

Haltinner, Kristin, and Dilshani Sarathchandra. "Considering Attitudinal Uncertainty in the Climate Change Skepticism Continuum." *Global Environmental Change* 68 (2021). https://doi.org/10.1016/j.gloenvcha.2021.102243.

Haltinner, Kristin, Dilshani Sarathchandra, Amber Ziegler, and Randolph P. Stuart. "It Wasn't Like a Big Light Bulb Moment": Factors that Contribute to Changing Minds on Climate Change. *Rural Sociology*. https://doi.org/10.1111/ruso.12460.

Haltinner, Kristin, Dilshani Sarathchandra, and Thomas Ptak. "How Believing that Climate Change Is a Conspiracy Affects Skeptics' Environmental Attitudes." *Environment: Science and Policy for Sustainable Development* 63, no. 3 (2021): 25–33.

Haltinner, Kristin, and Dilshani Sarathchandra. "Predictors of Pro-Environmental Beliefs, Behaviors, and Policy Support Among Climate Change Skeptics." *Social Currents* 9, no. 2 (2021): 180–202.

Haltinner, Kristin, and Dilshani Sarathchandra. "Pro-Environmental Views of Climate Skeptics." *Contexts* 19, no. 1 (2020): 36–41.

Haluza-DeLay, Randolph. "Religion and Climate Change: Varieties in Viewpoints and Practices." *Wiley Interdisciplinary Reviews: Climate Change* 5, no. 2 (2014): 261–79.

Hamilton, Lawrence C. "Did the Arctic Ice Recover? Demographics of True and False Climate Facts." *Weather, Climate, and Society* 4, no. 4 (2012): 236–49.

Hamilton, Lawrence C., Matthew J. Cutler, and Andrew Schaefer. "Public Knowledge and Concern about Polar-Region Warming." *Polar Geography* 35, no. 2 (2012): 155–68.

Hamilton, Lawrence C., Joel Hartter, and Kei Saito. "Trust in Scientists on Climate Change and Vaccines." *Sage Open* 5, no. 3 (2015). https://doi.org/10.1177/2158244015602752.

Harlos, Christian, Tim C. Edgell, and Johan Hollander. "No Evidence of Publication Bias in Climate Change Science." *Climatic Change* 140, no. 3–4 (2017): 375–85.

Harris, Dan, and Christine Brouwer. "Climate Scientists Claim 'McCarthy-Like Threats,' Say They Face Intimidation, Ominous E-Mails." *ABC News,* May 23, 2010. https://abcnews.go.com/WN/Media/climate-scientists-threat-global-warming-proponents-face-intimidation/story?id=10723932.

Hart, P. Sol, and Erik C. Nisbet. "Boomerang Effects in Science Communication:

How Motivated Reasoning and Identity Cues Amplify Opinion Polarization about Climate Mitigation Policies." *Communication Research* 39, no. 6 (2012): 701–23.
Harvey, Fiona. "World Is Running Out of Time on Climate, Experts Warn." *Guardian*, November 9, 2020. https://www.theguardian.com/environment/2020/nov/09/world-is-running-out-of-time-on-climate-experts-warn.
Harvey, Jeffrey A., Daphne Van Den Berg, Jacintha Ellers, Remko Kampen, Thomas W. Crowther, Peter Roessingh, and Bart Verheggen. "Internet Blogs, Polar Bears, and Climate-Change Denial by Proxy." *BioScience* 68, no. 4 (2018): 281–87.
Hasler, Béatrice S., Gilad Hirschberger, Tal Shani-Sherman, and Doron A. Friedman. "Virtual Peacemakers: Mimicry Increases Empathy in Simulated Contact with Virtual Outgroup Members." *Cyberpsychology, Behavior, and Social Networking* 17, no. 12 (2014): 766–71.
Hayhoe, Katharine. "Kathrine Hayhoe: Climate Scientist." Accessed November 17, 2020. http://www.katharinehayhoe.com/wp2016/biography/.
Hestres, Luis. "Fighting Climate Change Denial in the United States." In *Climate Change Denial and Public Relations*, edited by Nuria Almiron and Jordi Xifra, 217–32. New York: Routledge, 2020.
Hmielowski, Jay D., Lauren Feldman, Teresa A. Myers, Anthony Leiserowitz, and Edward Maibach. "An Attack on Science? Media Use, Trust in Scientists, and Perceptions of Global Warming." *Public Understanding of Science* 23, no. 7 (2014): 866–83.
Hoffmann, Roman, Raya Muttarak, Jonas Peisker, and Piero Stanig. "Climate Change Experiences Raise Environmental Concerns and Promote Green Voting." *Nature Climate Change* 12, no. 2 (2022): 148–55.
Hogg, Michael, and Dominic Abrams. *Social Identifications: A Social Psychology of Intergroup Relations and Group Processes.* London: Routledge, 1988.
Hogg, Michael, Deborah J. Terry, and Katherine M. White. "A Tale of Two Theories: A Critical Comparison of Identity Theory with Social Identity Theory." *Social Psychology Quarterly* 58, no. 4 (1995): 255–69.
Hornsey, Matthew J., and Kelly S. Fielding. "Attitude Roots and Jiu Jitsu Persuasion: Understanding and Overcoming the Motivated Rejection of Science." *American Psychologist* 72, no. 5 (2017): 459–73.
Howe, Peter D., Matto Mildenberger, Jennifer R. Marlon, and Anthony Leiserowitz. "Geographic Variation in Opinions on Climate Change at State and Local Scales in the USA." *Nature Climate Change* 5, no. 6 (2015): 596–603.
Hudson, John. "Energy Independence Is a Farce." *Atlantic*, June 30, 2012. https://www.theatlantic.com/business/archive/2012/06/energy-independence-is-a-farce/259253.
Idaho Department of Agriculture. "Always Growing." Accessed March 23, 2020. https://agri.idaho.gov/main/about/about-idaho-agriculture/.
Idaho Department of Environmental Quality. "Safe Pharmaceuticals Disposal."

Accessed March 27, 2020. https://www.deq.idaho.gov/pollution-prevention/p2-for-citizens/safe-pharmaceuticals-disposal/.

Iyengar, Shanto, and Kyu S. Hahn. "Red Media, Blue Media: Evidence of Ideological Selectivity in Media Use." *Journal of Communication* 59, no. 1 (2009): 19–39.

Iyengar, Shanto, Gaurav Sood, and Yphtach Lelkes. "Affect, Not Ideology: A Social Identity Perspective on Polarization." *Public Opinion Quarterly* 76, no. 3 (2012): 405–31.

Jasny, Lorien, and Dana R. Fisher. "Echo Chambers in Climate Science." *Environmental Research Communications* (2019). https://doi.org/10.1088/2515-7620/ab491c.

Jelen, Ted G., and Linda A. Lockett. "Religion, Partisanship, and Attitudes Toward Science Policy." *Sage Open* 4, no. 1 (2014): 2158244013518932.

Johnson, Dan R. "Transportation Into a Story Increases Empathy, Prosocial Behavior, and Perceptual Bias Toward Fearful Expressions." *Personality and Individual Differences* 52, no. 2 (2012): 150–55.

Johnson, Dan R. "Transportation Into Literary Fiction Reduces Prejudice Against and Increases Empathy For Arab-Muslims." *Scientific Study of Literature* 3, no. 1 (2013): 77–92.

Joireman, Jeff, Heather Barnes Truelove, and Blythe Duell. "Effect of Outdoor Temperature, Heat Pprimes an Anchoring on Belief in Global Warming." *Journal of Environmental Psychology* 30, no. 4 (2010): 358–67.

Jones, Jeffrey M. "Conservatives Greatly Outnumber Liberals in 19 U.S. States." Gallup, February 22, 2019. https://news.gallup.com/poll/247016/conservatives-greatly-outnumber-liberals-states.aspx.

Jones, Jeffrey M. "Democratic, Republican Confidence in Science Diverges." Gallup, July 16, 2021. https://news.gallup.com/poll/352397/democratic-republican-confidence-science-diverges.aspx.

Kahan, Dan M., Donald Braman, John Gastil, Paul Slovic, and C. K. Mertz. "Culture and Identity-Protective Cognition: Explaining the White-Male Effect in Risk Perception." *Journal of Empirical Legal Studies* 4, no. 3 (2007): 465–505.

Kearns, Laurel. "Noah's Ark Goes to Washington: A Profile of Evangelical Environmentalism." *Social Compass* 44, no. 3 (1997): 349–66.

Keller, Fred "America's Energy Independence Has Contributed to the Great American Comeback." The Hill, February 5, 2020. https://thehill.com/blogs/congress-blog/energy-environment/481679-americas-energy-independence-has-contributed-to-the.

Kessler, Glenn, Salvador Rizzo, and Meg Kelly. "President Trump Made 16,241 False or Misleading Claims in His First Three Years." *Washington Post,* January 20, 2020. https://www.washingtonpost.com/politics/2020/01/20/president-trump-made-16241-false-or-misleading-claims-his-first-three-years/.

Kimmel, Michael. *Angry White Men: American Masculinity at the End of an Era.* Paris: Hachette UK, 2017.

Kinney, Chad A., Edward T. Furlong, Stephen L. Werner, and Jeffery D. Cahill. "Presence and Distribution of Wastewater-Derived Pharmaceuticals in Soil Irrigated with Reclaimed Water." *Environmental Toxicology and Chemistry: An International Journal* 25, no. 2 (2006): 317–26.

Krause, Neal, and R. David Hayward. "Emotional Expressiveness During Worship Services and Life Satisfaction: Assessing the Influence of Race and Religious Affiliation." *Mental Health, Religion & Culture* 16, no. 8 (2013): 813–31.

Lahsen, Myanna. "Anatomy of Dissent: A Cultural Analysis of Climate Skepticism." *American Behavioral Scientist* 57, no. 6 (2013): 732–53.

Lahsen, Myanna. "Experiences of Modernity in the Greenhouse: A Cultural Analysis of a Physicist 'Trio' Supporting the Backlash against Global Warming." *Global Environmental Change* 18, no. 1 (2008): 204–19.

Langin, Katie. "Millions of Americans Drink Potentially Unsafe Tap Water: How Does Your Country Stack Up?" *Science*, February 12, 2018. https://www.sciencemag.org/news/2018/02/millions-americans-drink-potentially-unsafe-tap-water-how-does-your-county-stack.

Lee, Erika. *The Making of Asian America*. New York: Simon and Schuster, 2015.

Leiser, David, and Pascal Wagner-Egger. "Determinants of Belief—and Unbelief—in Climate Change." In *Climate of the Middle: Understanding Climate Change as a Common Challenge*, edited by Arjen Siegmann, 23–32. Berlin: Springer Briefs in Climate Studies, 2022.

Leiserowitz, Anthony. "American Risk Perceptions: Is Climate Change Dangerous?" *Risk Analysis* 25, no. 6 (2005): 1433–42.

Leiserowitz, Anthony A., Edward W. Maibach, Seth Rosenthal, John Kotcher, Parrish Bergquist, Matthew Ballew, Matthew Goldberg, and Abel Gustafson. "Climate Change in the American Mind: April 2019." Yale Program on Climate Change Communication. New Haven, CT: Yale University and George Mason University, 2019.

Leiserowitz, Anthony A., Edward W. Maibach, Connie Roser-Renouf, Nicholas Smith, and Erica Dawson. "Climategate, Public Opinion, and the Loss of Trust." *American Behavioral Scientist* 57, no. 6 (2013): 818–37.

Leiserowitz, Anthony, Connie Roser-Renouf, Jennifer Marlon, and Edward Maibach. "Global Warming's Six Americas: A Review and Recommendations for Climate Change Communication." *Current Opinion in Behavioral Sciences* 42 (2021): 97–103.

Lewandowsky, Stephan. "Climate Change Disinformation and How to Combat It." *Annual Review of Public Health* 42 (2021): 1–21.

Lewandowsky, Stephan, and John Cook. *The Conspiracy Theory Handbook*. Accessed August 17, 2020. https://www.climatechangecommunication.org/wp-content/uploads/2020/03/ConspiracyTheoryHandbook.pdf.

Lewandowsky, Stephan, Gilles E. Gignac, and Klaus Oberauer. "The Robust Relationship between Conspiracism and Denial of (Climate) Science." *Psychological Science* 26, no. 5 (2015): 667–70.

Lewandowsky, Stephan, Klaus Oberauer, and Gilles E. Gignac. "NASA Faked the Moon Landing—Therefore, (Climate) Science Is a Hoax: An Anatomy of the Motivated Rejection of Science." *Psychological Science* 24, no. 5 (2013): 622–33.

Li, Ya. "Nostalgia Promoting Pro-Social Behavior and Its Psychological Mechanism." *Open Journal of Social Sciences* 3 (2015): 177–86.

Lipka, Michael, and Benjamin Wormald. "How Religious Is Your State?" Pew Research Center. Accessed February 29, 2016. https://www.pewresearch.org/fact-tank/2016/02/29/how-religious-is-your-state/?state=alabama.

Liu, Marian. "The Coronavirus and the Long History of Using Diseases to Justify Xenophobia." *Washington Post,* February 13, 2020. https://www.washingtonpost.com/nation/2020/02/14/coronavirus-long-history-blaming-the-other-public-health-crises/.

Lohan, Tara. "Can California's Iconic Redwoods Survive Climate Change?" *The Revelator*, February 13, 2019. https://therevelator.org/redwoods-climate-change/.

Lowe, Lisa. *Immigrant Acts.* Durham, NC: Duke University Press, 1996.

Luker, Kristin. *Salsa Dancing in the Social Sciences.* Cambridge, MA: Harvard University Press, 2008.

MacInnis, Bo, and Jon Krosnick. "Climate Insights 2020: Partisan Divide." Resources for the Future. Accessed December 20, 2020. https://rff.org/publications/reports/climateinsights2020-partisan-divide/.

Mackenzie, Jillian. "Air Pollution: Everything You Need to Know." National Resource Defense Council. Accessed June 22, 2021. https://www.nrdc.org/stories/air-pollution-everything-you-need-know.

Mahaffy, Kimberly A. "Cognitive Dissonance and Its Resolution: A Study of Lesbian Christians." *Journal for the Scientific Study of Religion* 35, no 4. (1996): 392–402.

Malka, Ariel, Jon A. Krosnick, and Gary Langer. "The Association of Knowledge with Concern about Global Warming: Trusted Information Sources Shape Public Thinking." *Risk Analysis: An International Journal* 29, no. 5 (2009): 633–47.

Manjoo, Farhad. *True Enough: Learning to Live in a Post-Fact Society*. Hoboken, NJ: Wiley, 2008.

Marble, William, Salma Mousa, and Alexandra A. Siegel. "Can Exposure to Celebrities Reduce Prejudice? The Effect of Mohamed Salah on Islamophobic Behaviors and Attitudes." *American Political Science Review* 115, no. 4 (2021): 1111–28.

Markowitz, Ezra M., and Azim F. Shariff. "Climate Change and Moral Judgement." *Nature Climate Change* 2, no. 4 (2012): 243–47.

Matthews, Karly. "ACC Campus President to Embark on Electric Election 2020 Roadtrip." American Conservation Coalition. Accessed September 3, 2020. https://www.acc.eco/blog/2020/9/3/electric-election-announcement.

Mayden, Kelley D. "Peer Review: Publication's Gold Standard." *Journal of the Advanced Practitioner in Oncology* 3, no. 2 (2012): 117–22.

McCall, George J., and Jerry Laird Simmons. *Identities and Interactions*. New York: Free Press, 1978.

McCright, Aaron M. "Anti-Reflexivity and Climate Change Skepticism in the U.S. General Public." *Human Ecology Review* 22, no. 2 (2016): 77–107.

McCright, Aaron M., Katherine Dentzman, Meghan Charters, and Thomas Dietz. "The Influence of Political Ideology on Trust in Science." *Environmental Research Letters* 8, no. 4 (2013): loc. 044029.

McCright, Aaron M., and Riley E. Dunlap. "Anti-Reflexivity." *Theory, Culture & Society* 27, no. 2–3 (2010): 100–133.

McCright, Aaron M., and Riley E. Dunlap. "Bringing Ideology In: The Conservative White Male Effect on Worry about Environmental Problems in the USA." *Journal of Risk Research* 16, no. 2 (2013): 211–26.

McCright, Aaron M., and Riley E. Dunlap. "Cool Dudes: The Denial of Climate Change among Conservative White Males in the United States." *Global Environmental Change* 21, no. 4 (2011): 1163–72.

McCright, Aaron M., and Riley E. Dunlap. "The Politicization of Climate Change and Polarization in the American Public's Views of Global Warming, 2001–2010." *Sociological Quarterly* 52, no. 2 (2011): 155–94.

McDonald, Rachel I., Hui Yi Chai, and Ben R. Newell. "Personal Experience and the 'Psychological Distance' of Climate Change: An Integrative Review." *Journal of Environmental Psychology* 44 (2015): 109–18.

McGarty, Craig, Ana-Maria Bliuc, Emmaf Thomas, and Renata Bongiorno. "Collective Action as the Material Expression of Opinion-Based Group Membership." *Journal of Social Issues* 65, no. 4 (2009): 839–57.

McGrath, Matt. "Climate Change: IPCC Report Is 'Code Red for Humanity.'" *BBC News,* August 9, 2021. https://www.bbc.com/news/science-environment-58130705.

McKimmie, Blake M., Deborah J. Terry, Michael A. Hogg, Antony S. R. Manstead, Russell Spears, and Bertjan Doosje. "I'm a Hypocrite, But So Is Everyone Else: Group Support and the Reduction of Cognitive Dissonance." *Group Dynamics: Theory, Research, and Practice* 7, no. 3 (2003): 214–24.

Melley, Brian. "Explosive California Wildfires Could Burn into December." Associated Press, August 20, 2021. https://apnews.com/article/fires-environment-and-nature-california-18d949a81e52a85198e1089f711aae5c.

Messner, Michael A. *Politics of Masculinities: Men in Movements*. Lanham, MD: Altamira, 1997.

Motta, Matthew. "Changing Minds or Changing Samples? Disentangling Microlevel Stability and Macrolevel Growth in Anthropogenic Climate Change Beliefs." *International Journal of Public Opinion Research* 33, no. 2 (2021): 477–89.

Mullainathan, Sendhil, and Andrei Shleifer. "Media Bias." National Bureau of Economic Research Working Paper 9295, 2002. http://www.nber.org/papers/w9295.

Myers, Teresa A., Matthew C. Nisbet, Edward W. Maibach, and Anthony A. Leiserowitz. "A Public Health Frame Arouses Hopeful Emotions about Climate Change." *Climatic Change* 113, no. 3 (2012): 1105–12.

Nadelson, Louis, Cheryl Jorcyk, Dazhi Yang, Mary Jarratt Smith, Sam Matson, Ken Cornell, and Virginia Husting. "I Just Don't Trust Them: The Development and Validation of an Assessment Instrument to Measure Trust in Science and Scientists." *School Science and Mathematics* 114, no. 2 (2014): 76–86.

NASA (National Aeronautics and Space Administration). "Global Climate Change: Vital Signs of the Planet." Accessed December 23, 2020. https://climate.nasa.gov/vital-signs/carbon-dioxide/.

NASA (National Aeronautics and Space Administration). "Global Warming vs. Climate Change." Accessed December 23, 2020. https://climate.nasa.gov/global-warming-vs-climate-change/.

National Parks Conservation Association. "New Poll of Likely Voters Finds Unity in Public Support for National Parks." Accessed August 7, 2012. https://www.npca.org/articles/693-new-poll-of-likely-voters-finds-unity-in-public-support-for-national-parks.

Newheiser, Anna-Kaisa, Miguel Farias, and Nicole Tausch. "The Functional Nature of Conspiracy Beliefs: Examining the Underpinnings of Belief in the Da Vinci Code Conspiracy." *Personality and Individual Differences* 51, no. 8 (2011): 1007–11.

Nisbet, Matthew C. "Communicating Climate Change: Why Frames Matter for Public Engagement." *Environment: Science and policy for sustainable development* 51, no. 2 (2009): 12–23.

NOAA (National Oceanic and Atmospheric Administration). "Climate Models." Accessed November 17, 2021. https://www.climate.gov/maps-data/primer/climate-models.

NOAA (National Oceanic and Atmospheric Administration). "How Do We Study Past Climates?" Accessed November 27, 2021. https://www.climate.gov/maps-data/primer/past-climate.

Norgaard, Kari. *Living in Denial: Climate Change, Emotions, and Everyday Life.* Cambridge, MA: MIT Press, 2011.

"North China Floods Kill 15, 3 Missing." *Shanghai Daily,* October 13, 2021. https://archive.shine.cn/nation/North-China-floods-kill-15-3-missing/shdaily.shtml.

Northey, Gavin, Rebecca Dolan, Jane Etheridge, Felix Septianto, and Patrick Van Esch. "LGBTQ Imagery in Advertising: How Viewers' Political Ideology Shapes Their Emotional Response to Gender and Sexuality in Advertisements." *Journal of Advertising Research* 60, no. 2 (2020): 222–36.

NRCS (National Resources Conservation Service) Idaho. "Soil Health." Accessed November 20, 2021. https://www.nrcs.usda.gov/wps/portal/nrcs/main/id/soils/health/.

NREL (National Renewable Energy Laboratory). "Renewable Electricity Futures

Study." Accessed November 27, 2021. https://www.ucsusa.org/resources/benefits-renewable-energy-use#references.

NSF (National Science Foundation) National Science Board. "The State of U.S. Science and Engineering 2020." Accessed November 15, 2021. https://ncses.nsf.gov/pubs/nsb20201/invention-innovation-and-perceptions-of-science.

The Ocean Cleanup. "The Great Pacific Garbage Patch." Accessed November 27, 2021. https://theoceancleanup.com/great-pacific-garbage-patch.

Olson, Carin M., Drummond Rennie, Deborah Cook, Kay Dickersin, Annette Flanagin, Joseph W. Hogan, Qi Zhu, Jennifer Reiling, and Brian Pace. "Publication Bias in Editorial Decision Making." *JAMA* 287, no. 21 (2002): 2825–28.

Oppenheimer, Michael, Naomi Oreskes, Dale Jamieson, Keynyn Brysse, Jessica O'Reilly, Matthew Shindell, and Milena Wazeck. *Discerning Experts*. Chicago: University of Chicago Press, 2019.

The Oregonian. "Oregon Standoff Timeline: 41 Days of the Malheur Refuge Occupation and the Aftermath." January 8, 2019. https://www.oregonlive.com/portland/2017/02/oregon_standoff_timeline_41_da.html.

Oreskes, Naomi, and Erik M. Conway. *Merchants of Doubt: How a Handful of Scientists Obscured the Truth on Issues from Tobacco Smoke to Global Warming*. London: Bloomsbury, 2011.

Ornstein, Charles, and Katie Thomas. "Memorial Sloan Kettering Leaders Violated Conflict-of-Interest Rules, Report Finds." *New York Times*, April 4, 2019. https://www.nytimes.com/2019/04/04/health/memorial-sloan-kettering-conflicts-.html.

Painter, James, and Teresa Ashe. "Cross-National Comparison of the Presence of Climate Scepticism in the Print Media in Six Countries, 2007–10." *Environmental Research Letters* 7, no. 4 (2012). https://doi.org/10.1088/1748-9326/7/4/044005.

Pearce, Lisa D., and Arland Thornton. "Religious Identity and Family Ideologies in the Transition to Adulthood." *Journal of Marriage and Family* 69, no. 5 (2007): 1227–43.

Pellow, David, and Jasmine Vazin. "The Intersection of Race, Immigration Status, and Environmental Justice." *Sustainability* 11, no. 14 (2019): 3942–59.

Pereira, Andrea, and Jay Van Bavel. "Identity Concerns Drive Belief in Fake News." Unpublished manuscript, 2019. Available at: https://psyarxiv.com/7vc5d/.

Pilgeram, Ryanne. *Pushed Out: Contested Development and Rural Gentrification in the US West*. Seattle: University of Washington Press, 2021.

Poortinga, Wouter, Alexa Spence, Lorraine Whitmarsh, Stuart Capstick, and Nick F. Pidgeon. "Uncertain Climate: An Investigation into Public Scepticism about Anthropogenic Climate Change." *Global Environmental Change* 21, no. 3 (2011): 1015–24.

Popovich, Nadja, Livia Albeck-Ripka, and Kendra Pierre-Louis. "The Trump Administration Is Reversing 100 Environmental Rules. Here's the Full List." *New York*

Times, July 15, 2020. https://www.nytimes.com/interactive/2020/climate/trump-environment-rollbacks.html.

Powell, James Lawrence. *The Inquisition of Climate Science.* New York: Columbia University Press, 2011.

Prati, Gabriele, and Bruna Zani. "The Effect of the Fukushima Nuclear Accident on Risk Perception, Antinuclear Behavioral Intentions, Attitude, Trust, Environmental Beliefs, and Values." *Environment and Behavior* 45, no. 6 (2012): 782–98.

Price, Kiley. "IPCC Report: Climate Change Could Soon Outpace Humanity's Ability to Adapt." *Conservation International,* February 28, 2022. https://www.conservation.org/blog/ipcc-report-climate-change-could-soon-outpace-humanitys-ability-to-adapt.

Priest, Susanna Hornig. "Misplaced Faith: Communication Variables as Predictors of Encouragement for Biotechnology Development." *Science Communication* 23, no. 2 (2001): 97–110.

Quandt, Thorsten, Lena Frischlich, Svenja Boberg, and Tim Schatto-Eckrodt. "Fake News." *International Encyclopedia of Journalism Studies* (2019). https://doi.org/10.1002/9781118841570.iejs0128.

Quinn, Diane M. "Concealable Versus Conspicuous Stigmatized Identities." In *Stigma and Group Inequality*, edited By S. Levin and C. van Laar, 91–118. Mahwah, NJ: Psychology Press, 2006.

Ragan, Charles. *The Comparative Method.* Berkeley: University of California Press, 1987.

Rahmstorf, Stefan. "The Climate Sceptics." *Weather Catastrophes and Climate Change* (2004): 76–83. http://www.pik-potsdam.de/~stefan/Publications/Other/rahmstorf_climate_sceptics_2004.pdf.

Randall, Rosemary. "Loss and Climate Change: The Cost of Parallel Narratives." *Ecopsychology* 1, no. 3 (2009): 118–29.

Ray, Sarah Jaquette. *A Field Guide to Climate Anxiety*. Berkeley: University of California Press, 2020.

RepublicEn. "Are You One of Us?" Accessed December 19, 2021. https://republicen.org/about.

Rees, Jonas, Sabine Klug, and Sebastian Bamberg. "Guilty Conscience: Motivating Pro-Environmental Behavior by Inducing Negative Moral Emotions." *Climatic Change* 130 (2015): 439–52.

Ring, Wendy. "Inspire Hope, Not Fear: Communicating Effectively about Climate Change and Health." *Annals of Global Health* 81, no. 3 (2015): 410–16.

Roberts, David. "Did that New York Magazine Climate Story Freak You Out? Good." Vox, July 11, 2017. https://www.vox.com/energy-and-environment/2017/7/11/15950966/climate-change-doom-journalism.

Roeser, Sabine. "Risk Communication, Public Engagement, and Climate Change: A Role for Emotions." *Risk Analysis: An International Journal* 32, no. 6 (2012): 1033–40.

Rosa, Claudio D., Christiana Cabicieri Profice, and Silvia Collado. "Nature Experiences and Adults' Self-Reported Pro-Environmental Behaviors: The Role of Connectedness to Nature and Childhood Nature Experiences." *Frontiers in Psychology* 9 (2018). https://doi.org/10.3389/fpsyg.2018/01055.

Saad, Lydia. "Half in U.S. Are Now Concerned Global Warming Believers." Gallup, March 27, 2017. https://news.gallup.com/poll/207119/half-concerned-global-warming-believers.aspx.

Sarathchandra, Dilshani, and Kristin Haltinner. "How Believing Climate Change Is a 'Hoax' Shapes Climate Skepticism in the United States." *Environmental Sociology* 7, no. 3 (2021): 225–38.

Sarathchandra, Dilshani, and Kristin Haltinner. "A Survey Instrument to Measure Skeptics' (Dis)Trust in Climate Science." *Climate* 9, no. 2 (2021). https://doi.org/10.3390/cli9020018.

Sarathchandra, Dilshani, and Kristin Haltinner. "Trust/Distrust Judgments and Perceptions of Climate Science: A Research Note on Skeptics' Rationalizations." *Public Understanding of Science* 29, no. 1 (2020): 53–60.

Sarathchandra, Dilshani, Kristin Haltinner, and Matthew Grindal. "Climate Skeptics' Identity Construction and (Dis)Trust in Science in the United States." *Environmental Sociology* 8, no. 1 (2022): 25–40.

Sarathchandra, Dilshani, and Aaron M. McCright. "The Effects of Media Coverage of Scientific Retractions on Risk Perceptions." *Sage Open* 7, no. 2 (2017). https://doi.org/2158244017709324.

Sarathchandra, Dilshani, and Toby A. Ten Eyck. "To Tell the Truth: Keys in Newspaper Portrayals of the Public during Food Scares." *Food, Culture & Society* 16, no. 1 (2013): 107–24.

Schmid-Petri, Hannah, Silke Adam, Ivo Schmucki, and Thomas Häussler. "A Changing Climate of Skepticism: The Factors Shaping Climate Change Coverage in the US Press." *Public Understanding of Science* 26, no. 4 (2017): 498–513.

Schulman, Jeremy. "Every Insane Thing Donald Trump Has Said About Global Warming." *Mother Jones.* Accessed September 30, 2020. https://www.motherjones.com/environment/2016/12/trump-climate-timeline/.

Schutte, Nicola S., and Emma J. Stilinović. "Facilitating Empathy Through Virtual Reality." *Motivation and Emotion* 41, no. 6 (2017): 708–12.

Schwartz, Daniel, and George Loewenstein. "The Chill of the Moment: Emotions and Pro-Environmental Behavior." *Journal of Public Policy & Marketing* 36, no. 2 (2017): 255–68.

Seideman, David. *Showdown at Opal Creek: The Battle for America's Last Wilderness.* New York: Carroll & Graf, 1993.

Shan, Rachel. "Can Virtual Reality Drive Sustainable Behavior?" *Medium,* July 14, 2020. https://sustainabilityx.co/can-virtual-reality-save-the-environment-1129f76eae88.

Shapiro, Emily, and Neal Karlinsky. "Militia, Along with Family of Cliven Bundy, Take Over Federal Land at National Wildelife Refuge in Oregon." *ABC News*, January 3, 2016. https://abcnews.go.com/US/militia-family-cliven-bundy-federal-land-national-wildlife/story?id=36064909.

Sherkat, Darren E., and Christopher G. Ellison. "Structuring the Religion-Environment Connection: Identifying Religious Influences on Environmental Concern and Activism." *Journal for the Scientific Study of Religion* 46, no. 1 (2007): 71–85.

Shipman, Claire, and Katty Kay. *Womenomics.* New York: Harper Business, 2009.

Slovic, Paul, Melissa L. Finucane, Ellen Peters, and Donald G. MacGregor. "The Affect Heuristic." *European Journal of Operational Research* 177, no. 3 (2007): 1333–52.

Smith, Allan. "Republicans Who Believe in Climate Change Seek Alternative to Green New Deal." *NBC News*, March 10, 2019. https://www.nbcnews.com/politics/congress/republicans-who-believe-climate-change-seek-antidote-green-new-deal-n973146.

Smith, Nicholas, and Anthony Leiserowitz. "The Role of Emotion in Global Warming Policy Support and Opposition." *Risk Analysis* 34, no. 5 (2014): 937–48.

Sneed, Annie. "Ask the Experts: Does Rising CO2 Benefit Plants?" *Scientific American*, January 23, 2018. https://www.scientificamerican.com/article/ask-the-experts-does-rising-co2-benefit-plants1/.

Sobel, David. "Climate Change Meets Ecophobia." *Connect Magazine* 21, no. 2 (2007): 14–21.

Solar Reviews. "How Much Do Solar Panels Cost in Idaho, 2020?" Accessed December 15, 2020. https://www.solarreviews.com/solar-panel-cost/idaho.

Stafford, Marc, and Richard Scott. "Stigma Deviance and Social Control: Some Conceptual Issues." In *The Dilemma of Difference*, edited by S. C. Ainlay, G. Becker, and L. M. Coleman, 77–91. New York: Plenum, 1986.

Stanley, Samantha K., Teaghan L. Hogg, Zoe Leviston, and Iain Walker. "From Anger to Action: Differential Impacts of Eco-Anxiety, Eco-Depression, and Eco-Anger on Climate Action and Wellbeing." *Journal of Climate Change and Health* 1 (2021). https://doi.org/10.1016/j.joclim.2021.100003.

Stets, Jan E., and Peter J. Burke. "A Sociological Approach to Self and Identity." In *Handbook of Self and Identity,* edited by M. R. Leary and J. P. Tangney, 128–52. New York: Guilford, 2003.

Stets, Jan E., and Peter J. Burke. "Identity Theory and Social Identity Theory." *Social Psychology Quarterly* 63, no. 3 (2000): 224–37.

Stets, Jan E., and Richard T. Serpe, "Identity Theory." In *Handbook of Social Psychology*, 31–60. Dordrecht: Springer, 2013.

Stets, Jan E., and Teresa M. Tsushima. "Negative Emotion and Coping Responses Within Identity Control Theory." *Social Psychology Quarterly* 64, no. 3 (2001): 283–95.

Stryker, Sheldon. "The Interplay of Affect and Identity: Exploring the Relationships

of Social Structure, Social Interaction, Self, and Emotion." Panel presentation at American Sociological Association Conference, Chicago, 1987.
Stuart, Toby E., and Waverly W. Ding. "When Do Scientists Become Entrepreneurs? The Social Structural Antecedents of Commercial Activity in the Academic Life Sciences." *American Journal of Sociology* 112, no. 1 (2006): 97–144.
Suldovsky, Brianne. "The Information Deficit Model and Climate Change Communication." Oxford Research Encyclopedia of Climate Science Online, Accessed August 14 2021, 2017. https://oxfordre.com/climatescience/view/10.1093/acrefore/9780190228620.001.0001/acrefore-9780190228620-e-301.
Swami, Viren, Rebecca Coles, Stefan Stieger, Jakob Pietschnig, Adrian Furnham, Sherry Rehim, and Martin Voracek. "Conspiracist Ideation in Britain and Austria: Evidence of a Monological Belief System and Associations between Individual Psychological Differences and Real-World and Fictitious Conspiracy Theories." *British Journal of Psychology* 102, no. 3 (2011): 443–63.
Swami, Viren, and Adrian Furnham. "Political Paranoia and Conspiracy Theories." In *Power, Politics, and Paranoia: Why People are Suspicious of Their Leaders*, edited by J. Prooijen and P. Lange, 218–36. Cambridge, MA: Cambridge University Press. 2014.
Taber, Charles S., and Milton Lodge. "Motivated Skepticism in the Evaluation of Political Beliefs." *American Journal of Political Science* 50, no. 3 (2006): 755–69.
Tajfel, Henri, ed. *Differentiation Between Social Groups: Studies in the Social Psychology of Intergroup Relations*. Cambridge, MA: Academic Press, 1978.
Tajfel, Henri, and John Turner. "Social Identity Theory of Intergroup Behavior." In *Psychology of Intergroup Relations,* edited by Stephen Worchel and William Austin, 7–24. Chicago: Nelson-Hall, 1986.
Tarlach, Gemma. "The Five Mass Extinctions that Have Swept Our Planet." *Discover,* July 18, 2018. https://www.discovermagazine.com/the-sciences/mass-extinctions.
Taylor, Cedric, dir. *Nor Any Drop to Drink.* PBS, 2018. https://www.noranydropfilm.com.
Taylor, Laura K., and Catherine Glen. "From Empathy to Action: Can Enhancing Host-Society Children's Empathy Promote Positive Attitudes and Prosocial Behavior Toward Refugees?" *Journal of Community & Applied Social Psychology* 30, no. 2 (2020): 214–26.
Thoits, Peggy. "Multiple Identities: Examining Gender and Marital Status Differences in Distress." *American Sociology Review* 51 (1986): 259–72.
Tingting, Deng. "In China, the Water You Drink Is as Dangerous as the Air You Breathe." *Guardian,* June 2, 2017. https://www.theguardian.com/global-development-professionals-network/2017/jun/02/china-water-dangerous-pollution-greenpeace.
Truelove, Heather Barnes, and Jeff Joireman. "Understanding the Relationship between Christian Orthodoxy and Environmentalism: The Mediating Role of Perceived Environmental Consequences." *Environment and Behavior* 41, no. 6 (2009): 806–20.

Turner, John, and Howard Giles. *Intergroup Behavior.* Oxford, UK: Blackwell, 1981.

Turner, John C., Michael A. Hogg, Penelope J. Oakes, Stephen D. Reicher, and Margaret S. Wetherell. *Rediscovering the Social Group: A Self-Categorization Theory.* Oxford, UK: Blackwell, 1987.

Union of Concerned Scientists. "Benefits of Renewable Energy Use." Accessed October 10, 2020. https://www.ucsusa.org/resources/benefits-renewable-energy-use.

Union of Concerned Scientists. "Coal Power Impacts." Accessed July 9, 2019. https://www.ucsusa.org/resources/coal-power-impacts

United Nations Environment Programme. "Air Pollution and Climate Change: Two Sides of the Same Coin." October 11, 2020. https://www.unenvironment.org/news-and-stories/story/air-pollution-and-climate-change-two-sides-same-coin.

Uscinski, Joseph E., Karen Douglas, and Stephan Lewandowsky. "Climate Change Conspiracy Theories." Oxford Research Encyclopedia of Climate Science Online. Accessed November 21, 2017. https://oxfordre.com/climatescience/climatescience/abstract/10.1093/acrefore/9780190228620.001.0001/acrefore-9780190228620-e-328.

US Energy Information Administration. "Fossil Fuels Continue to Account for the Largest Share of U.S. Energy." September 18, 2019. https://www.eia.gov/todayinenergy/detail.php?id=41353.

Van Bavel, Jay. "Group Think." *Hidden Brain.* National Public Radio, September 20, 2021. https://hiddenbrain.org/podcast/group-think/.

Van Bavel, Jay, and Dominic Packer. *The Power of Us: Harnessing Our Shared Identities to Improve Performance and Promote Social Harmony.* New York: Little, Brown Spark, 2021.

Wachinger, Gisela, Ortwin Renn, Chloe Begg, and Christian Kuhlicke. "The Risk Perception Paradox—Implications for Governance and Communication of Natural Hazards." *Risk Analysis* 33, no. 6 (2013): 1049–65.

Wang, Jaesun, and Seoyong Kim. "Analysis of the Impact of Values and Perception on Climate Change Skepticism and Its Implication for Public Policy." *Climate* 6, no. 4 (2018). https://doi.org/10.3390/cli6040099.

Wang, Susan, Zoe Leviston, Mark Hurlstone, Carmen Lawrence, and Iain Walker. "Emotions Predict Policy Support: Why It Matters How People Feel about Climate Change." *Global Environmental Change* 50 (May 2018): 25–40.

Watts, Jonathan. "Climatologist Michael E. Mann: 'Good People Fall Victim to Doomism: I Do Too Sometimes." *Guardian,* February 27, 2021. https://www.theguardian.com/environment/2021/feb/27/climatologist-michael-e-mann-doomism-climate-crisis-interview.

Weed, Mike. "Capturing the Essence of Grounded Theory: The Importance of Understanding Commonalities and Variants." *Qualitative Research in Sport, Exercise and Health* 9, no. 1 (2017): 149–56.

White, Lynn. "The Historical Roots of Our Ecologic Crisis." *Science* 155, no. 3767 (1967): 1203–7.
The White House. "The Build Back Better Framework." Accessed April 5, 2022. https://www.whitehouse.gov/build-back-better/.
The White House. "Fact Sheet: The Recovery Act Made." Accessed February 25, 2016. https://obamawhitehouse.archives.gov/the-press-office/2016/02/25/fact-sheet-recovery-act-made-largest-single-investment-clean-energy.
Yale Climate Opinion Maps, 2021. Accessed April 5, 2022. https://climatecommunication.yale.edu/visualizations-data/ycom-us/.
Yueh, Chieling. "Classification of Conversations: Distinguishing between Opposing Climate Change Communities on Reddit." Bachelor's thesis, Utrecht University, 2021.
Zhou, Jack. "Boomerangs versus Javelins: How Polarization Constrains Communication on Climate Change." *Environmental Politics* 25, no. 5 (2016): 788–811.
Zhou, Li. "Trump's Racist References to the Coronavirus Are His Latest Effort to Stoke Xenophobia." Vox, June 23, 2020. https://www.vox.com/2020/6/23/21300332/trump-coronavirus-racism-asian-americans.
Zielinski, Sarah. "Climate Change Will Accelerate Earth's Sixth Mass Extinction." *Smithsonian Magazine,* April 30, 2015. https://www.smithsonianmag.com/science-nature/climate-change-will-accelerate-earths-sixth-mass-extinction-180955138/.
Žižek, Slavoj. *The Sublime Object of Ideology*. London: Verso, 2019.

index

Ingram Content Group UK Ltd.
Milton Keynes UK
UKHW041038100723
424752UK00002BA/39